The Greenwich Guides to Astronomy

The *Greenwich Guides* are a series of books on astronomy for the beginner. Each volume stands on its own but together they provide a complete introduction to the night sky, everything it contains, and how astronomers are discovering its secrets. Written by experts from the Old Royal Observatory at Greenwich, they are right up to date with the latest information from space exploration and research and are suitable for observers in both the northern and southern hemisphere.

Available now
The Greenwich Guide to Stargazing
The Greenwich Guide to The Planets

To be published
The Greenwich Guide to Stars, Galaxies and Nebulae
The Greenwich Guide to Astronomy in Action

The Old Royal Observatory, Greenwich, London is open daily to visitors. It is the home of Greenwich Mean Time and the Greenwich Meridian which divides East from West. It also houses the largest refracting telescope in Great Britain.
 For more information write to: Marketing Department, National Maritime Museum, Greenwich, London SE10 9NF.

The Greenwich Guide to
The Planets

Stuart Malin

George Philip
in association with
The National Maritime Museum, London

This book is copyrighted under the Berne Convention. All rights reserved. Apart from any fair dealing for the purpose of private study, research, criticism or review, as permitted under the Copyright Act, 1956, no part of this publication may be reproduced, stored in a retrieval system, or transmitted in any form or by any means, electronic, electrical, chemical, mechanical, optical, photocopying, recording, or otherwise, without prior written permission. All enquiries should be addressed to the Publishers.

British Library Cataloguing in Publication Data

Malin, S.R.C.
 The Greenwich Guide to the planets.
 1. Planets
 I. Title II. Greenwich, Royal Observatory
 523.4 QB601

ISBN 0-540-01128-2

©The Trustees of The National Maritime Museum 1987
First published by George Philip,
27A Floral Street, London WC2E 9DP

Printed in Hong Kong

Acknowledgements

I am grateful to many people for their assistance in the preparation of this book. In particular I would like to thank Carole Stott, Marilla Fletcher and other colleagues at the Old Royal Observatory, Greenwich, David Hughes of Sheffield University and Lydia Greeves of George Philip.

I am also grateful to the following for permission to reproduce their illustrations: R. J. Allenby, Jr., p. 46; European Space Agency, p. 45; Federation of Astronomical Societies, p. 31 (Bob Tuffnell), p. 37 (top) (Geoff Pearce), p. 57 (David Strange); John Guest, p. 44; David Hughes, p. 47; H. V. Keller, Max Planck-Institut für Aeronomie, Lindau/Harz, West Germany p. 90; Patrick Moore, p. 84; NASA/JPL, pp. 20, 33, 34, 35, 40, 41, 52 (top), 52 (bottom), 54, 55 (left), 55 (right), 59, 60, 61, 62, 63, 64, 65, 67, 69, 70, 71, 72, 73, 76, 77, 78, 79, 80; The Trustees, The National Gallery, London, p. 39; The Trustees, The National Maritime Museum, London, pp. 12 (top left), 12 (top right), 14 (bottom), 18, 75 (left), 89, pp. 2, 26, 50 (Jim Stevenson); Royal Astronomical Society, London, pp. 22–25, 27 (left), 27 (right), 28, 82, 85; Royal Greenwich Observatory, Herstmonceux, pp. 9, 19, 51 (top), 51 (bottom).

Jacket illustrations: NASA/JPL (front); Old Royal Observatory, Jim Stevenson (back). The back jacket illustration shows the Old Royal Observatory, Greenwich.

FRONTISPIECE *Full Moon rising over Flamsteed House, Greenwich, the home of the Astronomers Royal from 1676 to 1948.*

Contents

Introduction 6
1 The Solar System 7
2 The Sun and the Moon 18
3 Mercury 30
4 Venus 36
5 Earth 42
6 Mars 48
7 Jupiter 56
8 Saturn 66
9 Uranus, Neptune and Pluto 74
10 Interplanetary Debris 86
11 Are We Alone? 93
Index 94

Introduction

For as long as man has looked at the sky, he has been fascinated by the planets and their strange movements. Much has been learnt about these bodies by looking at them with the naked eye or through telescopes, but our knowledge has been revolutionized in recent years as a result of the information obtained from space probes.

This book is an introduction to the Sun and its planets. After looking at the Solar System as a family, each of its members is examined in detail, telling you what you can see from Earth and giving you the latest information from satellites and space probes. Some of the most intriguing bodies are not the planets, but the moons that orbit them or the comets and asteroids which cross their paths. The final chapter explores the question of whether we are alone and speculates on the origins of life and the existence of planetary systems elsewhere.

1 · The Solar System

On a cosmic scale, the planets amount to almost nothing. They are a set of small balls that happen to orbit the Sun, and only show up at all because they reflect sunlight. The Sun itself is a very ordinary star. If we lived in the neighbourhood of the next nearest star, the Sun's planets would be insignificant from that distance and it is unlikely that we would know that they even existed. Nevertheless, the planets are of great interest to us for several reasons. Firstly, as we are so close to them they appear as some of the brightest objects in the sky; secondly, because they are our nearest neighbours in space; thirdly, because we live on one of them; fourthly ... the list could go on and on.

Before discussing the individual planets in detail, it is useful to get an overall picture of the Solar System and that is the purpose of this first chapter. (The Solar System is the collective name for the Sun, planets, moons and a few other odds and ends that drift around the Sun.)

The word 'planet' means a wanderer. Back in the days before electric light the only lights our ancestors would see if they went outdoors at night would be from the Moon and stars, so they became very familiar with the star-patterns. Although the stars move across the sky during the night (because of the rotation of the Earth), the relative positions of the stars do not change so the pattern they form remains constant. However, there are a few rather bright 'stars' that do change their positions over a period of days or months, and this property made them stand out as being of particular interest. These 'wandering stars' are the planets.

In the beginning

About 15 thousand million years ago the Universe suddenly came into existence with a mighty bang. Just at that instant, everything in the Universe was in the same place, but the explosion heated it to an incredible temperature and flung it out into space. The effects of this explosion are still with us: the Universe is still expanding and there is a faint background radiation that is left over from the initial searing heat. Another Greenwich Guide describes how the material from the Big Bang formed itself into galaxies and stars but here we are concerned with the Sun and planets. They probably came into existence as follows.

If we were to have visited the region of the Solar System some 4600 million years ago, we would have found a gently rotating cloud of gas and dust, mostly made up of hydrogen but also containing a small proportion of other elements. This is the cloud from which the Sun and planets condensed. The cloud was not uniform, but had regions of higher and lower density, which tended to get more extreme as the gravity of the denser regions pulled in more material, making the less dense regions even emptier. In this way our cloud broke up into a large number of blobs which moved around, bumping into one another and sometimes breaking up and sometimes coalescing. The bigger the blob, the greater its gravitational attraction, so the bigger blobs tended to absorb the smaller ones and the whole mix got lumpier. It might seem that the ultimate outcome of all this would be one large blob—the

Sun—but the fact that the original gas cloud was rotating had an important effect on what happened.

When a rotating body gets compressed, it rotates faster. This can be seen in the bathtub. The bath water can appear to be almost stationary, but when it gets squeezed into the plug-hole it spins like a whirlpool. The same will have happened to our blobs; as they collected more material they spun faster. If all the gas cloud had tried to form a single star, it would have spun so fast that it would have flown apart. So instead of forming a ball, the gravitational forces compressed the cloud to a flat spinning disc, somewhat like a discus, but made up of millions of freely-moving lumps. Within the disc, the force of gravity trying to pull all the lumps towards the centre was balanced by the effect of rotation, which acted to keep them spread out.

As well as gravity and rotation, there is now a third factor to be considered—temperature. Whenever a new particle joined a lump by crashing into it, it changed its energy into heat, so compression was accompanied by a rise in temperature. The greatest compression was towards the centre of the disc, and here a huge ball of hydrogen got extremely hot as gravity compressed it into a smaller ball. When the density and temperature reached a critical point, some of the hydrogen atoms fused to form helium and released a great deal of energy. This was a thermonuclear reaction, like a hydrogen bomb, and once it got under way the great ball of gas became a star. The star was, of course, the Sun, but at this stage it was only a proto-sun; nothing like so hot or so compact as it is today. Even so it will have set up a strong temperature gradient in the disc; hot in the middle and cooler towards the edges. The Sun, too, has had problems with its spin, but it has somehow managed to solve them very well because it is now rotating very sedately once every 27 days. It probably slowed down by throwing out gases as a strong wind sweeping through the disc. We will now leave the Sun to settle down to its present form and concentrate on the disc.

The disc material has now become a large number of lumps up to 1000 kilometres across, all orbiting the Sun in the same direction. They will continue to collide with one another to form a few big lumps and these will sweep up most of the smaller ones. When they get big enough, these primitive planets can also hold on to some of the lighter gases to form an atmosphere. Their success in doing this will depend on several things. First, they need to be big enough. Really large lumps, like those that were to become Jupiter and Saturn, could hold on to almost anything that came their way. Smaller lumps could hold on to only the cooler (and hence slower-moving) gases and so those towards the edge of the disc, further from the heat of the Sun, were likely to be most successful. The outer planets were also favoured because much of the lighter gas would have been swept away from the vicinity of the Sun by the wind.

As they collected material and compressed it through their own gravity, the planets would get hotter. So hot that they might melt and allow the heavier materials to sink to the centre and the lighter ones to rise. Ice and other volatile material evaporated into the atmosphere, some to escape into space and some to remain with the planet and eventually condense back onto the surface when the planet had cooled again; size and temperature would decide what was kept and what was lost.

Most of this process, from gas cloud to Sun and planets, took place in just a few million years, though the outer planets took rather longer to form. Since then, the planets have been steadily cooling and mopping up the debris that was left around. Studies of the Moon and planets show that most of the cratering occurred when they were still young, and that more recent craters tend to be smaller, so perhaps the mopping up procedure is nearly complete.

Early ideas

The five brightest planets were well known to the ancients and have been given the names of Gods: Mercury, Venus, Mars, Jupiter and Saturn. To these they added the Sun and Moon to make up a set of seven. These are probably the 'seven stars in the sky' in the old song 'Green grow the rushes, oh' and are certainly the basis of the days of the week. Saturday, Sunday and Monday come from Saturn, Sun and Moon, but the remaining days in English are named after Anglo-Saxon equivalents of the Roman gods. To see their planetary origins, we have to go to another language, such as Italian. Starting with Monday, the Italian days are *lunedi* (lune = Moon), *martedi* (Mars), *mercoledi* (Mercury), *giovedi* (Jove = Jupiter) and *venerdi* (Venus). Similar associations can be traced in other languages.

The names given to the planets were chosen to be the most appropriate to the behaviour or appearance of the planet. The god Mercury is the winged messenger, always on the move; similarly, the planet Mercury is the most rapidly moving and most elusive of the planets, frequently heralding the appearance of the Sun in the dawn sky, or chasing after it at dusk. Saturn, the slowest moving planet, bears the name of a more sedate and gloomy god (though he had his moments—a saturnalia is a drunken orgy). Mars is the god of war, and the planet has a reddish appearance, suggesting blood. Venus is the goddess of beauty and no one could deny that the planet is very beautiful, shining bright in the morning or evening sky. Jupiter is the top god, so it is appropriate that his planet spends most of its time dominating the night sky.

For centuries, the movements of the planets amongst the stars could not be explained, and our superstitious forebears thought they had mystical properties which could influence our lives. In those days there was no distinction between astronomy and astrology, and the men who studied the stars and planets, though they could not explain them, claimed to be able to interpret their meanings in terms of horoscopes. People were supposed to come under the influence of planets, which would confer on them the qualities of the gods after whom they were named. For example, someone might be mercurial, or another might have a saturnian temperament. This sort of thing still goes on, of course, and most people enjoy reading 'their stars', but nowadays only for entertainment.

Towards an understanding

As well as making a living out of fortune-telling, the old stargazers built up a great deal of information about how the planets moved.

Illustration from the German astronomer Johannes Kepler's Cosmographical Enigma *(1596). This was an early attempt to explain the distances of planets from the Sun by placing their orbits on spheres nested with the five regular solids (cube, tetrahedron, dodecahedron, icosahedron and octahedron).*

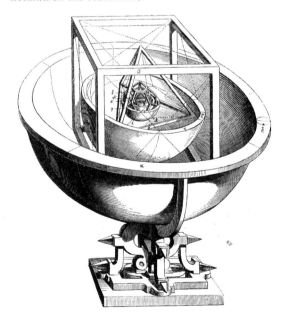

Mercury is the fastest moving and never strays far from the Sun, appearing first on one side of it, then on the other and coming back to where it started every 116 days. The movement of Venus against the stars is the next fastest. The remaining planets go more slowly and can appear in the night sky long after sunset, though they always remain near the path of the Sun amongst the stars. Observations of this kind required some sort of explanation, so the philosophers got to work.

Some of the early ideas about planetary motions had little to do with science. At one time it was thought that the movements of the planets (and, indeed, all movements) were controlled by angels. This led to quite serious arguments about how many angels could sit on the head of a pin. A more productive theory had the Earth at the centre of a series of concentric transparent spheres. These crystalline spheres rotated independently, the nearer ones each carrying one planet (including the Sun and Moon), and the most distant one carrying the stars. Although this makes a pretty enough picture, it does not account for the observed motions of the planets unless the spheres rotate in a most erratic way.

One great problem with the movements of the planets is the so-called *retrograde motion* shown by Mars, Jupiter and Saturn, a phenomenon that was well known to the early astronomers. Most of the time these planets move fairly steadily from west to east against the star background, but sometimes they stop and go back towards the west before again moving east. The first man to develop a theory to account for this was Ptolemy, in the second century AD. He proposed a series of circles centred on the Earth, representing the average orbits of the planets, but suggested a planet did not move steadily along its circle. Instead, it went round and round a smaller circle, the centre of which moved steadily along the big circle. The big circle was called the *deferent* and the smaller one was an *epicycle*. By choosing appropriate sizes for the epicycles and deferents, and an appropriate rate of movement along them, he was able to account for the observed movements of the planets quite well, including the retrograde motion. But the Ptolemaic theory did not even try to explain why the planets should choose to follow this curious path.

Light dawns

The great breakthrough came more than a thousand years later when a Polish astronomer, Nicholas Copernicus, made the audacious proposal that the planets orbited around the Sun, not the Earth. By abandoning the old geocentric (Earth-centred) hypothesis and drawing the planetary orbits as circles around the Sun (the heliocentric hypothesis), Copernicus was able to represent the planetary motions just as well as Ptolemy, but without the need for any epicycles. The one 'planet' that remained circling the Earth was the Moon, so the theory recognized its essential difference from the other planets. The theory also recognized for the first time the special place of the Sun as the centre of the Solar System and the very un-special place of the Earth as just another planet, circling the Sun like all the others.

Although the Copernican theory presented a simple and elegant picture of how the planets moved, in other respects it was no better than the Ptolemaic theory, since it made no attempt to explain what caused the planets to move in the way they did. And it still did not fit the observations any better.

It was the observations of Galileo in about 1610 that finally led to the abandonment of the

ABOVE RIGHT *Ptolemy's theory of planetary motions with the Earth at the centre and the Sun, Moon and planets moving around it on epicycles and deferents.*

BELOW RIGHT *Copernicus's theory of planetary motions with the Sun at the centre, the Earth and planets moving around it in circles, and the Moon circling the Earth. (Both diagrams are not to scale.)*

THE SOLAR SYSTEM

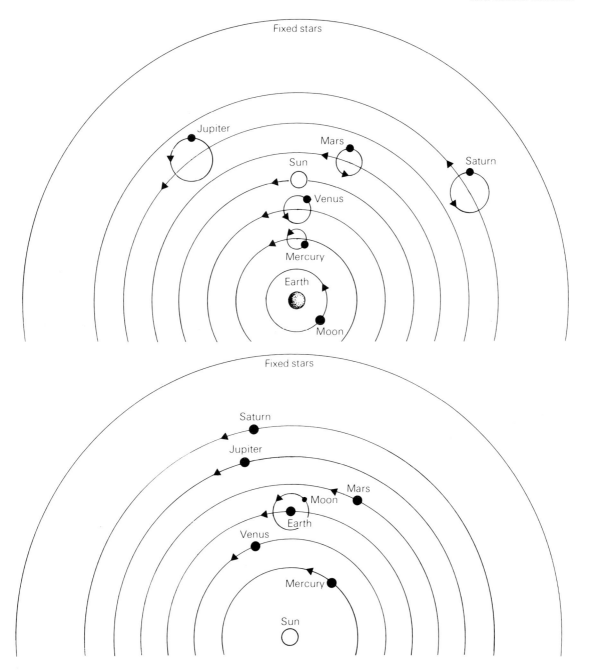

11

Nicholas Copernicus (1473–1543), the Polish astronomer who developed the heliocentric (Sun-centred) theory of planetary motions.

ABOVE Galileo Galilei (1564–1642), the Italian astronomer, physicist and mathematician whose observations showed the flaws in the geocentric theory of planetary motions.

BELOW The changing phases of Venus show that it is sometimes between the Earth and the Sun and sometimes beyond the Sun.

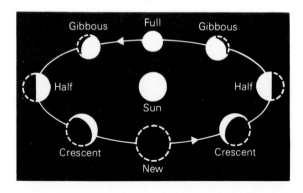

geocentric theory. With his newly-invented telescope he was able to see four moons orbiting around Jupiter, thus demonstrating that heavenly bodies could rotate around planets other than the Earth. He also saw that Venus was not just a point of light but goes through a series of phases, like the Moon, varying from a slim crescent to full disc. This means that Venus is sometimes between the Earth and the Sun (crescent phase) and sometimes beyond the Sun (full phase), so it must go round the Sun.

The work that led to our present ideas about how the planets move was a fine example of how a scientific investigation should be conducted, but seldom is. Ideally, one should first make careful observations of the phenomenon being investigated, then try to find some general rules that describe what is going on, and finally develop a theory that explains why the rules are followed. The careful observations were made by Tycho Brahé, a Danish astronomer with a private observatory on the island of Hven, in the strait between Denmark and Sweden. In 1599, Brahé took on Johannes Kepler as an assistant, who then devoted his life to trying to sort out some underlying patterns from the wealth of planetary observations. Kepler's achievements are summarized in his three laws which are of fundamental importance to the development of planetary astronomy.

Kepler's first law states that the planets move in ellipses, with the Sun at one focus. An ellipse can be drawn by putting a loop of thread loosely over two pins and moving a pencil round so that its point always keeps the thread taut. Each pin is at a focus of the ellipse. The shape of the ellipse can be varied by adjusting the distance between the pins: when they are very close together, the ellipse is nearly a circle, but it becomes almost a straight line as they are moved further apart. For most of the planets, the foci are so close together that the orbits are nearly circular, so Copernicus was not too badly wrong. Nevertheless, they are ellipses and it is interesting that the Sun is not even at the exact centre of the ellipse, but displaced towards one end. The first law describes the shapes of the orbits; Kepler's second law goes on to describe the rate at which the planets go round their orbits. It states that the line drawn from the Sun to a planet sweeps out equal areas in equal times. For this to be so, the planet has to move faster when it is nearer the Sun than it does when it's further away. Kepler's law does not explain what causes this speeding up and slowing down, but it describes exactly when, where and by how much the speed changes around the orbit.

The first two laws are obeyed separately by

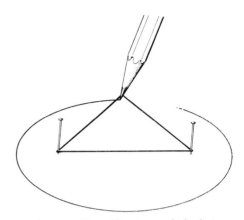

How to draw an ellipse. The pins mark the foci.

each of the planets. The third law, which was stated in 1618, nine years after the first two, relates all the planetary orbits to one another and makes it possible to draw a scale model of the Solar System. It relates the time taken for the completion of an orbit to the average distance from the Sun: the square of the time is proportional to the cube of the distance. This is best illustrated with an example. The Earth takes one year for an orbit and is at a distance of (say) 1 unit from the Sun. Following Kepler's law, Jupiter, which orbits the Sun in 11.9 years, must be at an average distance of 5.2 units, since 11.9×11.9 (the square of the time) $= 141.6 = 5.2 \times 5.2 \times 5.2$ (the cube of the distance). Similarly, knowing the orbital period of any planet, we can deduce its average distance from the Sun in terms of Earth-to-Sun units. How these units can be converted to miles or kilometres will be considered later.

Kepler's three laws fit the observations exactly. They are not an approximation to what goes on, like Ptolemy's epicycles or Copernicus's circles. They not only provide a compact way of summarizing all the observations of the planets, but can also be used to predict where a planet will be at any chosen time, past, present or future. Indeed, about the only question the three laws do not answer is why the planets move in this way. This final and most important part of the theory of planetary motion was left to Isaac Newton, arguably the greatest scientist who ever lived.

We all know the story of Newton being hit by a falling apple and blaming it on gravity. This deduction is not particularly clever—any one of us could have made it. Newton's spark of genius was to recognize that the force that makes apples fall to the ground is exactly the same force that keeps the Moon in orbit round the Earth. This realization was only the first step towards his theory. It took prodigious mathematical skill and application to sort out how gravity behaves and then to apply it to the orbits of the planets. Newton's law of gravity is simple and elegant when written mathematically ($F = G m_1 m_2/d^2$), but rather clumsy when written out in words: the force between two bodies is proportional to the product of their masses,* inversely proportional to the square of their separation, and acts along the line joining them. The force on each body is the same. The apple pulls the Earth with exactly the same force that the Earth pulls the apple; it is only because the Earth is so massive that the apple's force on it has no noticeable effect. In 1687 Newton published his results in his famous *Principia Mathematica*, in which he also showed that all Kepler's laws are a direct consequence of the law of gravity. Here at last was the one underlying principle that explained all the planetary motions. The key to the planetary orbits was the gravitational attraction of the Sun.

It would be satisfying to be able to say that that was the end of the story, but science is not like that. For over two centuries, Newton's law of gravity was thought to give an exact description of the planetary movements. But with increasingly accurate observations, some small but

*Mass is not the same as weight, because weight depends on gravity. However, even in space where there is no gravity and a body is weightless it still has mass, as you would soon find out if it collided with you.

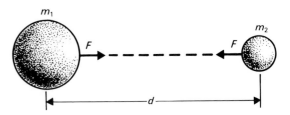

ABOVE *Newton's law of gravity. The force F is the same on each of the two bodies and is given by*
$$F = G\, m_1\, m_2/d^2,$$
where G is a constant, m_1 and m_2 are the masses of the bodies and d is the distance between them.

BELOW *Sir Isaac Newton (1642–1727), the English mathematician and physicist who developed the law of gravity and used it to explain the motions of the planets.*

significant departures from the predicted orbits were noticed. In particular, the point on the orbit of Mercury that is closest to the Sun does not stay fixed, but moves slowly around the Sun. Newton's theory can account for only part of this. At first the remaining part was thought to be due to the presence of an undiscovered planet still closer to the Sun, but careful searches failed to find it. The solution came when Albert Einstein developed his theory of relativity in the early years of this century. Einstein's theory is of great importance to physics in general, and is nearer to 'the truth' than that of Newton. But the difference between the predictions of the two theories is very small indeed for planetary orbits, and it is still customary to use Newton's law for all except the most refined of calculations.

The size of the Solar System

Newton's theory provides us with a scale model of the Solar System, which would work just as well if it were a million kilometres across, or a hundred million kilometres or a million million kilometres. The problem is to establish what the actual distances are. This can be avoided by defining the distance from the Earth to the Sun as one astronomical unit (1 a.u.) and measuring all the planetary distances in a.u.s, but it is clearly desirable to know what the distances are in terms of measurable units, such as kilometres. Once one measurement of a distance in the Solar System has been made in kilometres, we can deduce how many kilometres there are in 1 a.u. and then convert all the distances to kilometres. The measurement does not need to be of the distance from the Earth to the Sun; it can be from the Sun to another planet, or of the distance between two planets.

As they so often did, the Greeks were the first to come up with a solution. Aristarchus devised a geometrical method for comparing the Earth/Sun and Earth/Moon distances by measuring the angle between the Sun and the Moon when the latter was exactly half illuminated. Although the principle was right, the half-moon measurement was extremely difficult to make to the accuracy required, and the answer came out much too small. A much better geometrical method was devised by Edmond Halley (see Chapter 4). Nowadays, the most accurate method is to use radar and note the time it takes for a signal travelling at the (accurately known) speed of light to travel to a planet and back. In round numbers, the a.u. is 150 million kilometres.

The present picture

The Solar System is made up of the Sun, planets and various other bits of space debris that orbit the Sun, including comets, asteroids and meteors. The Sun accounts for over 99.9 per cent of the Solar System's mass. Moving outwards from the Sun the planets are: Mercury, Venus, Earth, Mars, Jupiter, Saturn, Uranus, Neptune, Pluto; it may help to remember the order with the sentence: My Very Early Morning Jam Sandwich Usually Nauseates People. The orbit of Pluto is the most elliptical, which sometimes brings it nearer to the Sun than Neptune.

It is difficult to visualize the size of the Solar System, but the following scale model may give some idea. Imagine the Sun as the dome of St Paul's Cathedral—a globe with a diameter of about 34 metres in the City of London. Then Mercury will be the size of a grapefruit about 1.5 kilometres away. Venus is represented by a football at a distance of 2.6 kilometres and the Earth is another football 3.6 kilometres from St Paul's. The last of the inner planets is Mars, somewhat larger than Mercury but smaller than the Earth and Venus, orbiting in a circle with a radius of 5.5 kilometres. There is then a gap before the largest planet of all, Jupiter, represented by a ball with a diameter of 3.5 metres—not much smaller than the cupola on top of St Paul's Cathedral. Its orbit has a radius of 19 kilometres, taking it through London's outer

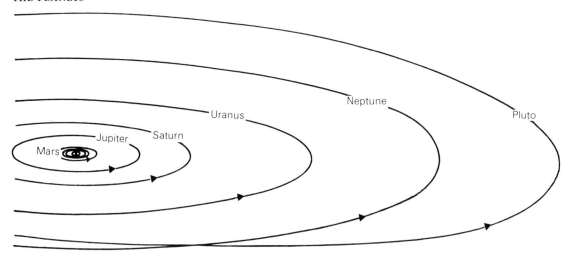

ABOVE *The relative sizes of the orbits of the planets round the Sun. At present Pluto is in the part of its orbit that comes between the orbits of Uranus and Neptune.*

BELOW *The relative sizes of the Sun and planets.*

suburbs. Saturn is somewhat smaller than Jupiter, corresponding to a 3-metre diameter ball at a distance of 35 kilometres. Uranus is smaller still, a very large beach ball with a diameter of 1.3 metres, orbiting just inland from Brighton in the south, and nearly as far as Bedford and Cambridge to the north. Neptune is only slightly smaller than Uranus, its 110-kilometre radius orbit taking it through Southampton and the Straits of Dover. Pluto is the smallest planet of all, a tennis ball that can come as close to London as Dover or go as far away as Dunkirk on the other side of the Channel. On this scale, with the Sun the size of the dome of St Paul's, all the planets could fit comfortably into a two-car garage. For completeness, the nearest star would be more than twice as far away as Australia.

The two smallest planets are those nearest to

and farthest away from the Sun—Mercury and Pluto respectively. The largest—Jupiter and Saturn—come in the middle with the intermediate planets filling in the gaps. The number of satellites per planet also peaks in the middle and declines towards and away from the Sun. The table summarizes some basic planetary statistics.

Planetary statistics
In each case the quantities are relative to 1 for the Earth.

	Distance from Sun	Diameter	Mass	Density	Length of year	Length of day	Number of moons
Mercury	0.39	0.38	0.06	0.98	0.24	58	0
Venus	0.72	0.95	0.82	0.95	0.61	243	0
Earth	1.00	1.00	1.00	1.00	1.00	1.00	1
Mars	1.52	0.53	0.11	0.71	1.88	1.03	2
Jupiter	5.20	11.27	318	0.24	11.86	0.41	16
Saturn	9.54	9.44	95	0.13	29.46	0.43	17
Uranus	19.2	4.10	15	0.22	84.02	0.65	15
Neptune	30.1	3.88	17	0.30	164.8	0.67	2
Pluto	39.4	0.17	0.002	0.24	248	6.4	1

2 · The Sun and the Moon

The Sun

Since we all know that the Sun is a star, what is it doing in a book about the planets? The reason is that the Sun is so important that it cannot be left out. To do so would be like talking about leaves, twigs and acorns without even mentioning the oak tree.

If you went out of your way to select a typical star from amongst all those in the firmament, you would probably end up with the Sun. It is neither young nor old, it is not unusually big or small and its temperature is about average. It is neither at the centre nor on the edge of our galaxy of stars. In fact, about the only extraordinary thing about

A 1582 woodcut illustrating the Sun and the Moon.

the Sun is its collection of planets. And perhaps even this is not so unusual. Since the Sun is so typical in every other respect, perhaps most stars have planets, and it is only because they are so difficult to detect that they have not been found elsewhere. We will return to this question later.

The other features that mark the Sun out from the other stars are its brightness and its apparent size—after all, you do not need to wear starglasses, or avoid exposing your skin to the night sky for fear of starstroke. But this is simply because we are only 150 million kilometres from the Sun. If you got that close to a hot star you would be frazzled, or if you got that close to the centre of a cooler giant star, you would actually be inside it! The next nearest star after the Sun is more than a quarter of a million times as far away from us as the Sun. No wonder the stars look so faint.

The Sun is a huge ball of gas—mostly hydrogen—which is glowing white-hot. It is hot because of the conversion of minute proportions of the hydrogen into helium with the release of large quantities of energy, just as in a hydrogen-bomb explosion, except that the process goes on continuously in the Sun. Most of the reaction takes place deep inside the Sun where the temperature is many millions of degrees Celsius. The surface is relatively cool—a mere 6000°C— and that is what gives the Sun its characteristic whitish-yellow colour; hotter stars appear blue and cooler ones only red hot. The surface is cooler than the centre because it is radiating heat and light away into space. A tiny proportion of this radiation—less than 0.00000005 per cent—is intercepted by the Earth, and similar amounts by the other planets.

The intensity of the radiation from the Sun diminishes rapidly with distance, so the radiation falling on a square kilometre of the surface of Mercury is 6.6 times as intense as that falling on a square kilometre of the Earth and 178 times as intense as that falling on the same area of Jupiter. It is not possible to estimate the temperature of a planet exactly from the amount of solar radiation it receives, as it also depends on factors such as internal heat sources, how much of the solar radiation is reflected and how good the planet is at radiating away heat, but solar radiation gives a guide. Obviously, Mercury will be a lot hotter than the Earth and Pluto is much colder.

Because the other stars are so remote, they appear only as points of light even through large telescopes, but the Sun is close enough for us to see features on its surface. The most obvious are sunspots, which can best be seen by projecting an image of the Sun on to a piece of white card held behind the eyepiece of a small telescope. *Never look directly at the Sun through a telescope or binoculars if you value your eyesight.* The spots are small and dark, often clustered in groups, the larger of which last for several months. Studying how they move across the Sun's disc has shown that the Sun rotates on its axis once every 27 days. The interesting thing is that all the planets travel round the Sun approximately above the

Sunspots on the face of the Sun.

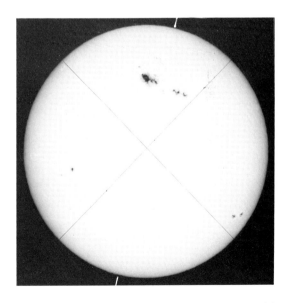

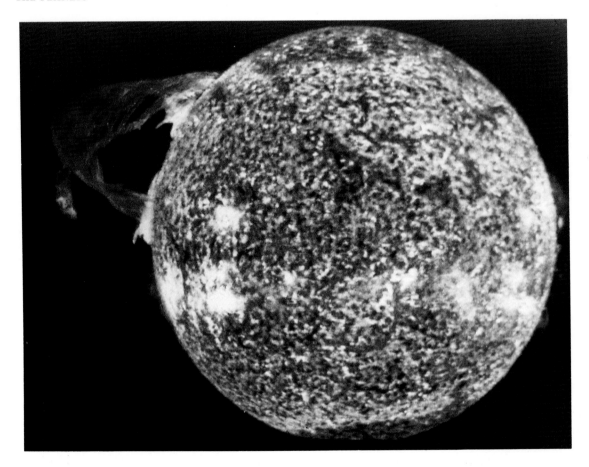

A large prominence on the Sun, 19 December 1973, photographed in hydrogen light to obtain more surface detail. This outburst is exceptional, but there is always some activity to be seen on the Sun.

Sun's equator and in the same direction as the Sun rotates. This shows that the planets really are associated with the Sun and were not just picked up as bits of dirt casually encountered by the Sun as it floated through space.

As well as the sunspots, it is sometimes possible to see prominences on the edge of the Sun's disc, though these are best seen through a special filter that transmits only hydrogen light. The prominences show up most clearly against the black sky background, but they can occur anywhere on the Sun. They are huge clouds of gas blasted out from the Sun's surface, mostly to fall back, but occasionally fired out so fast that they escape into space. Besides these more spectacular outbursts the Sun emits a continuous stream of electrically charged particles—the solar wind—which blows through the Solar System and interacts with a planet's magnetic field, if there is one.

So, as well as providing the gravitational

attraction that keeps the planets in their orbits, the Sun also provides nearly all their heat and light, and immerses them in a stream of electrically charged particles. On top of all this, the Sun also emits weak radio-waves and X-rays. It should now be clear that the planets cannot be considered as isolated bodies in space—they all come very directly under the Sun's influence.

The Moon

There is no need to apologize for including the Moon in a book about planets; although it is not itself a planet, it is intimately connected with one. This also applies to the moons of other planets, which are included in the chapter on the appropriate planet.

After the Sun, the Moon is the brightest object in the sky, but it has no light of its own and only shines because it is reflecting the light of the Sun. It is not even a very good reflector—about as good as a ploughed field—reflecting a mere 7 per cent of the light that falls on it. Even at full Moon, it sheds less than three millionths as much light as the Sun, and the fact that we can see by moonlight has more to do with the adaptability of our eyes than the brightness of the Moon. However, although we can see by moonlight, the low light levels are beyond the reach of the colour-sensitive detectors in our eyes and everything appears in shades of grey. This is also why the colours attributed to the stars and planets are not very convincing to the naked eye but show up better when the light is amplified by a telescope.

A lunation is the time it takes for the Moon to go through a cycle from new to first quarter (half Moon), to full to third quarter (half Moon again) and back to new. This takes $29\frac{1}{2}$ days, and is the original basis of the month ('moonth'). It is unfortunate that there are $12\frac{1}{3}$ lunations in a year rather than a round number and this has given rise to all sorts of complications in the calendar.

The Christian calendar stretches the month to 30 or 31 days to make it fit the year, but has to go back to the Moon to determine the date of Easter; the Moslems stick to twelve lunations and shrink the year so that the time of new year moves back slowly through the seasons; the Jewish calendar has 12 months in general, but sometimes the last month is repeated to keep it in phase with the year; the Ethiopians have 13 months, the last one being very short.

Features can be seen on the Moon even with the naked eye. There are a number of darker areas which show up best at full Moon and with a bit of imagination these can be seen as the face of the man in the Moon. The fact that he always appears the same and never turns his back on us shows that the Moon does not rotate relative to the Earth. (Viewed from space, the Moon would be seen to rotate once a month, but as this is the time the Moon takes to orbit the Earth, the same face is always towards us.)

As well as the dark areas, which early astronomers called *maria* because they thought they were seas, a keen-eyed observer can pick out a few brighter spots. These are craters and the lighter coloured material that surrounds them. But for a proper look at the craters you really need a telescope or a pair of binoculars. Through these the Moon is a truly magnificent sight, and even the most hardened professional astronomers still get a kick out of seeing it. It is best viewed along the *terminator*—the boundary between the dark and light sides of the Moon, where the craters cast long shadows in the setting Sun. The Moon is literally covered with craters, though they are rather sparse over the maria. The maria themselves are massive craters (note their circular outlines) that were sub-

OVERLEAF *Half Moon (pp. 22–3) and full Moon (pp. 24–5) showing maria, highlands and craters. The darker maria and lighter highlands show up well at full Moon, but the craters are best seen near the terminator on the half Moon picture.*

THE PLANETS

THE SUN AND THE MOON

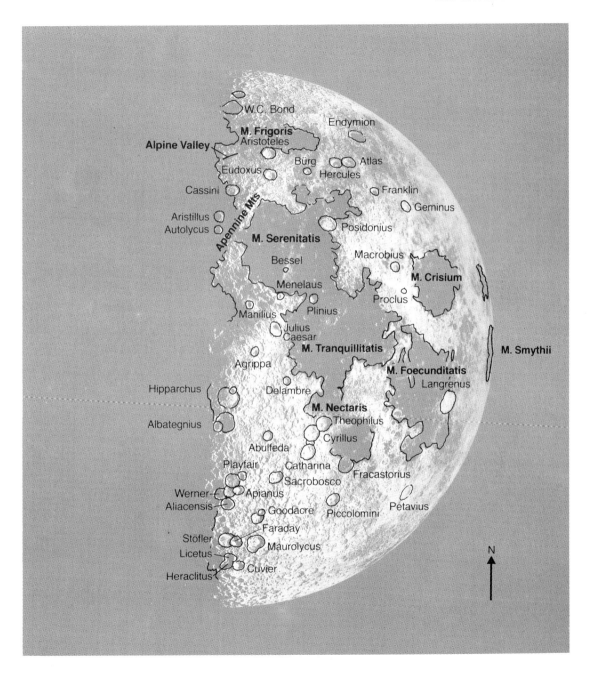

THE PLANETS

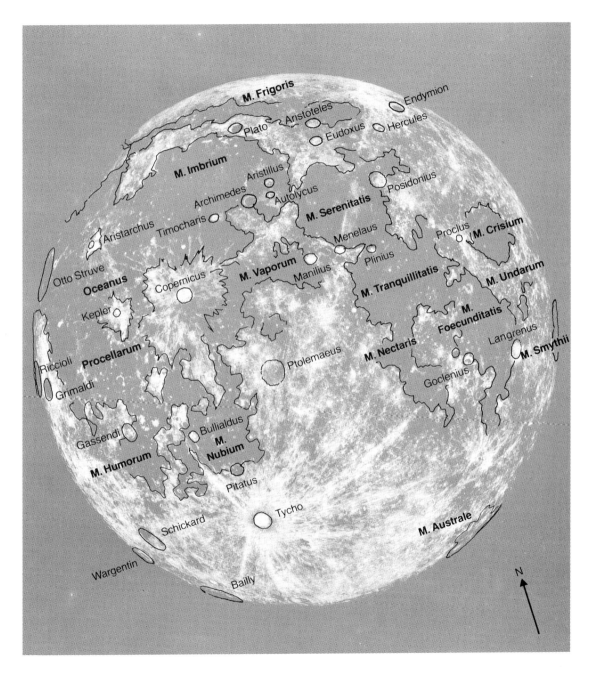

Full Moon over the dome of the 28-inch refracting telescope at the Old Royal Observatory, Greenwich.

sequently filled by lava-flows, and the smaller craters in them were caused by later impacts.

For a long time there was a fierce argument over whether the craters were of volcanic origin or were caused by collisions with meteorites, but this has now been resolved in favour of collisions. Certainly there is (or, at least, has been) volcanic activity on the Moon, or there would have been no lava-flows to form the maria, but, with a very few small exceptions the craters are the result of impacts. They were formed when a lump of rock (meteorite) travelling at thousands of kilometres an hour crashed on to the Moon. The heat created by the impact vaporized most of the meteorite and nearby parts of the lunar surface. The gas thus formed exploded like a huge bomb, sending out shock waves which gouged out a circular crater and threw the contents out onto the Moon's surface, sometimes to a distance of thousands of kilometres. The explosion also compressed the surface, particularly immediately under the point of impact, and this later relaxed, with the central point popping up higher than the surrounding area. Some of the larger rocks thrown out by the impact formed their own secondary craters around the first one when they crashed back onto the surface. Smaller impacts produced smaller craters, and those less than 20 kilometres in diameter (still large enough to swallow London) tend to be bowl-shaped, with no central peak.

Since the Earth is more or less in the same position as the Moon, why isn't it too covered in craters? It must have received the same sort of meteorite bombardment as the Moon. The reason is that there are processes at work on the Earth (see Chapter 5) that are constantly changing its surface, whereas the Moon retains all its old scars. By looking at the Moon we can deduce the sort of bombardment that has been suffered by the Earth in earlier times.

Unravelling the history of the Moon requires painstaking and detailed study. From this it

Galileo's drawings of the Moon as seen through his telescope.

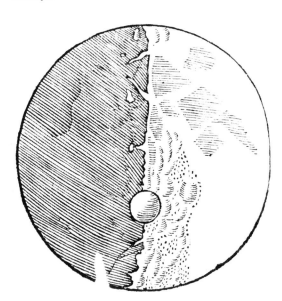

appears that the larger craters, including the maria basins, were formed about 4 billion years ago when there was still a lot of debris left over from the formation of the planets. The Moon itself was still young and for the next billion years there was still sufficient molten material near its surface to flow into the craters. Since then, the Moon has solidified and more recent craters have remained unfilled. Although the meteorite bombardment has eased off greatly, there is no reason to suppose that it has stopped. The fact that no new craters have been seen since observations of the Moon began is hardly surprising; the interval that has elapsed since the invention of the telescope is only an instant in the history of the Moon.

Our knowledge of the Moon has been greatly improved as a result of spacecraft exploration, firstly with orbiters (the first picture of the back of the Moon was taken by a Soviet spacecraft in 1959), then with unmanned landers, and certainly not least with direct exploration by man. One of the most exciting films ever made was from the 'Moon buggy', driving for the first time across the surface of a foreign astronomical body. Seismographs left on the Moon record moonquakes, measurements have been made of

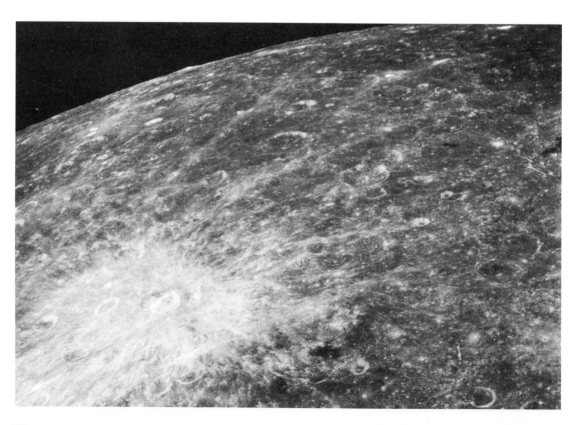

Part of the far side of the Moon, photographed from Apollo 8 in 1968.

heat-flow, reflectors on the Moon enable us to measure its distance from the Earth to within a metre, and so on. But the real bonus of the programme of Moon exploration is the samples of moon rock, over 300 kilogrammes of which have been brought back to Earth, for study and analysis in laboratories all over the world.

Eclipses

It is a matter of chance that the Sun and Moon appear to be the same size in the sky. Because of this quirk of fate, the Moon is just capable of covering the Sun when the two are perfectly aligned, and then it causes a total solar eclipse. This is a rare and beautiful phenomenon which people will travel a long way to see. It is usually necessary to travel, because a total eclipse can be seen from only a limited area on the Earth, and even then it lasts for only a few minutes; from other parts of the world a small crescent of the Sun is left uncovered. As the Moon encroaches on the Sun's disc, the sky gradually darkens. Then, when the last part of the Sun is obscured it is suddenly as dark as midnight with the stars shining. It is also possible to see the faint *corona*, or upper atmosphere of the Sun, surrounding the Moon's disc. (See *The Greenwich Guide to Stargazing* for a list of future eclipses.)

Obviously a total solar eclipse occurs at new Moon. At full Moon (again, only when the alignment is perfect), the Moon can pass through the Earth's shadow and this produces a lunar eclipse. These are more common than solar eclipses, because the Earth casts a bigger shadow than the Moon, but they are less spectacular. This is partly because it is dark anyway, and partly because the Moon can still be seen as a dull brown disc illuminated by the sunlight that has been bent by the Earth's atmosphere. A lunar eclipse can be seen equally well from anywhere on the night-time side of the Earth, so there is no need to travel. During a partial lunar eclipse, the Earth's shadow takes a bite out of the Moon's disc, but does not swallow it completely. A moonman who wanted to see the Sun totally eclipsed by the Earth during a partial eclipse would have to travel into the bite.

A total eclipse of the Sun can only be seen from the small area of Earth covered by the tip of the Moon's narrowing shadow. A partial eclipse will be seen over the region where the Moon's widening shadow is cast.

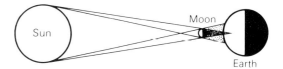

3 · Mercury

Orbital period: 88 days; *rotation period*: 59 days; *distance from Sun*: 70×10^6 km (maximum), 46×10^6 km (minimum); *diameter* 4880 km; mass 3.30×10^{23} kg; *density*: 5.4×10^3 kg m^{-3}*; *satellites*: none.

The view from Earth

Mercury is the nearest planet to the Sun. Viewed from the Earth it is never more than 28° (about the span of a hand at arm's length) from the Sun so it can only be seen close to the horizon within two hours of sunset or sunrise. Most of the time it appears even closer to the Sun and is very difficult to pick out in the glow of the rising or setting Sun even though, in favourable conditions, it can be as bright as Sirius (the brightest star). In Chapter 1, we said that it takes 116 days for Mercury to complete a cycle relative to the Sun, from one maximum eastern separation (*elongation* is the technical term) to the next. But the orbital period is, in fact, 88 days. The reason for the difference between these two figures is that the first is based on how Mercury appears when viewed from the Earth, which itself is moving round the Sun, whereas the second is the time it would take as viewed by an observer fixed in space. Thus, 116 days is the *synodic* period and 88 days the *sidereal* period. The latter is the more useful since it is a property of Mercury itself, whereas the synodic period also involves the Earth's orbit.

Like all planets, Mercury gives out no light of its own and can only be seen because of the sunlight it reflects. When it is on the far side of the Sun it appears full, i.e. with its whole disc illuminated, but then of course the Sun gets in the way. New Mercury occurs when the planet is between the Earth and the Sun, but again it is difficult to see except on the rare occasions when it is exactly on the Earth-Sun line and appears as a small spot on the Sun's disc. This is known as a *transit of Mercury*. When the planet is most easily viewed, i.e. when it is at maximum elongation, it is a 'half-Mercury'. However, it is not easy to see its shape even with a good telescope, partly because it is very small (about as big as a 1p coin at 600 metres) but also because it is viewed very low down in the sky where turbulence in the atmosphere spoils the image in much the same way as distant buildings appear unsteady and broken up when looked at through the hot air above a bonfire.

It *is* possible to see Mercury when it is high in the sky and there is less atmosphere in the way to spoil the image, but then the Sun is also in the sky and a telescope with accurate setting-circles is required to pick up the planet against the bright sky background. This should not be attempted by any but the most experienced observers because of the danger of permanently damaging your eyesight by inadvertently pointing the telescope at the Sun. A steady image of the planet can be

---

*3.3×10^{23} is a shorthand way of writing 330,000,000,000,000,000,000,000 i.e. $\times 10^{23}$ means 'move the decimal point 23 places to the right'. Similarly, $\times 10^{-5}$ means 'move the decimal point 5 places to the left' so $6.2 \times 10^{-5} = 0.000062$. The abbreviation kg m^{-3} stands for kilogrammes per cubic metre; the density of water is 1×10^3 kg m^{-3}.

Mercury and Venus at twilight. Mercury is the fainter, higher image.

seen in this way, although with only the vaguest hints of surface markings. This is unfortunate, because a planet's rate of rotation can be measured most easily from studies of the day-to-day movements of surface features. The difficulty of doing this with Mercury led to some misleading estimates of its rotation period. First, it was thought to have a 24-hour rotation period (like the Earth), and then Giovanni Schiaparelli made a series of observations about a hundred years ago from which he deduced a rotation period of 88 days. In other words, he thought the planet rotated about its axis in the same time that it took to complete an orbit round the Sun. This meant that it would always have the same side towards the Sun, just as our own Moon always has the same side towards the Earth. If this were so, the side exposed to the Sun would be hot enough to melt lead and the permanently dark side would be near to absolute zero, nearly 700°C colder.

When radio-wave measurements of the planet's temperature were made in the 1960s, no such contrast was found between the light and dark sides, so something must have been wrong with Schiaparelli's results. By bouncing radio-waves off a moving object and comparing the frequency of the return signal with that transmitted, it is possible to deduce how fast the object is approaching or receding. When this was done with opposite sides of Mercury, it was found that the period of rotation is 58.6 days, so the Sun *does* rise and set on Mercury, and that is why the day/night temperature difference is not too extreme.

It takes 58.6 days for Mercury to rotate once about its axis, but in that time it has also travelled two-thirds of the way round the Sun. If you were standing on the surface of Mercury, the Sun would appear to have gone only one-third of the way round the sky, so a 'day' (the interval between successive appearances of the Sun in the same part of the sky) would be 176 Earth-days, or two complete orbits.

Although Mercury appears as a bright spot in the sky, in fact it has a rather dark surface which, like our Moon, reflects only 7 per cent of the sunlight that falls on it. In comparison fresh snow reflects over 90 per cent of the light that falls on it and soot reflects almost nothing. Clearly, the surface of Mercury is more like the latter.

Mercury is also rather similar to the Moon in terms of size, which we can easily calculate from a knowledge of the planet's distance and the size it appears in the sky. Mercury has a diameter 1.4 times that of the Moon, though its volume is nearly three times that of the Moon.

Mercury has no moon of its own, which makes it difficult to measure the planet's mass. For a planet with a moon, all one has to do is measure the distance of the moon and the period of the moon's orbit and then deduce the planet's mass from a simple formula derived from Newton's law of gravity. In the case of Mercury, a less direct method involving studies of the effect of Mercury on the orbit of other bodies in the Solar System had to be used. Obviously, the nearer another body approaches Mercury, the greater the effect, so the object used was the minor planet Eros, whose orbit brings it close to Mercury. The mass of Mercury is about $4\frac{1}{2}$ times the mass of the Moon. Dividing the mass by the volume gives the average density of Mercury, which is definitely not like that of the Moon, but rather similar to that of the Earth.

The high density of Mercury means that, although the surface looks moonlike, the interior must be very different. Mercury probably has an iron core, like the Earth. From a knowledge of Mercury's mass and diameter, we can deduce the surface gravity, which turns out to be about a third of that on the Earth, and the *escape velocity*. When a ball is thrown in the air, it comes back down. The faster it leaves your hand, the longer it takes to fall back, and if it is thrown at a great enough speed, i.e. faster than the escape velocity, it will never come back. There is no-one on Mercury to throw balls, but the same principle would apply to gas on the surface of the planet, in other words to an atmosphere. The molecules of a gas all move about at different speeds but with an average speed that increases with temperature. Those moving faster than the escape velocity will disappear off into space. Because of Mercury's high temperature and low surface gravity, all but the heaviest of gases would have escaped from the planet long ago, so the conclusion is that Mercury has no atmosphere.

Spacecraft exploration

Everything said about Mercury so far has been deduced from observations made from the Earth, painstakingly built up over very many years. While these provide basic facts about the planet's orbit, size, mass etc., it has to be admitted that they don't tell us much about what Mercury is really like.

This situation was changed dramatically in 1974 when an American spaceprobe, *Mariner 10*, visited Mercury. Its route had taken it first to Venus, not just to look at that planet but mainly to make use of Venus's gravitational pull to modify the spacecraft's orbit and bring it towards Mercury at exactly the velocity required for the encounter. The modified path was chosen so that *Mariner 10* would orbit the Sun every 176 days, thus bringing it close to Mercury every two Mercurian years. The first encounter was on 29

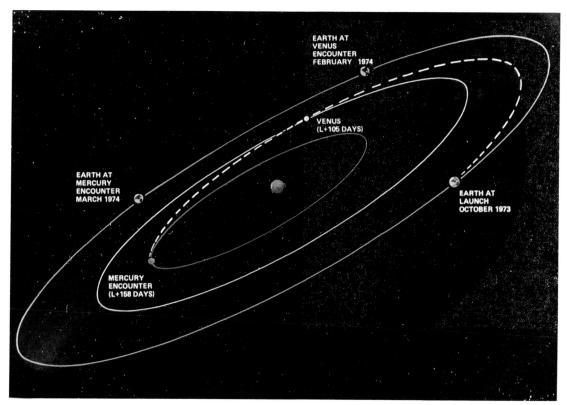

Mariner 10's route from Earth to Mercury, via Venus.

March 1974, at an altitude of 700 kilometres over the dark side of Mercury, and there were further encounters on 21 September 1974 and 16 March 1975. The second encounter was high over the sunlit side, but for the third encounter *Mariner 10* came right in to 327 kilometres—close enough to see features only 100 metres across. Earth gets no closer than a distance of 76 million kilometres and even then the atmosphere spoils the view.

The most spectacular results from the *Mariner 10* mission are the magnificent series of detailed television pictures of 40 per cent of the surface. What wouldn't Schiaparelli have given for a view like this? At first sight, the surface looks remarkably like that of the Moon, pitted with thousands of craters of all sizes, but there are important differences. The most obvious is the absence of the large, darker regions known as maria. More detailed examination shows that smaller craters on Mercury tend to have similar profiles to larger ones on the Moon, as can be explained because of the greater gravity on Mercury. For the same reason secondary craters, formed by lumps of rock thrown out from the larger craters, are closer in to the original crater than they would be on the Moon. Mercury also has features known as *scarps* that are not found on the Moon. These are long, low ridges that may have been caused by the wrinkling of the surface as the interior contracted.

The largest feature on Mercury is the Caloris

Mercury photographed by Mariner 10 *from a distance of 20,000 kilometres on 31 March 1974. The picture is made up as a mosaic from eighteen separate images. Craters with diameters of up to 200 kilometres can be seen, but the Caloris Basin is not included.*

Basin, an area ringed by mountains and riddled with cracks, which probably resulted from a huge impact about 4 billion years ago. So great was this impact that the reverberations from it passed through the centre of the planet and produced a 'weird terrain' consisting of a mass of rounded hills and valleys on the opposite side. Despite its 1300-kilometre diameter, the Caloris Basin was nearly missed because it occurs right at the edge of the photographed region, but fortunately there is enough of it visible to infer the whole feature.

One of the simpler tasks is the naming of the newly discovered features, except that there are so many of them! The recommendation laid down by the committee appointed by the International Astronomical Union was that craters should be named after authors, artists and musicians, valleys after radio-observatories and scarps after ships that explored the Earth. Plains were to be given the names used for Mercury in many different languages. Thus Discovery is a scarp, Arecibo is a valley and Shakespeare is one of the craters. Inevitably there are a few exceptions, such as Caloris itself.

Pictures were only a part of the information obtained by *Mariner 10*. Because the spaceprobe went so close to Mercury, its orbit was strongly affected by the planet's gravitational attraction and the deviation could be used to make a better estimate of Mercury's mass. The size of the planet was accurately measured by noting how long the radio signals from *Mariner 10* were interrupted while the spacecraft passed behind the planet. The rate at which the radio signals faded as the spacecraft went into eclipse and sensitive ultra-violet observations confirmed that there is just about no atmosphere on Mercury, though minute traces of helium gas were found.

The wealth of data from *Mariner 10* will keep planetary geologists happy for many years to come, sorting out the history of the bombardment that caused the craters (the newer ones overlap the older ones, and this helps to establish the sequence), interpreting the other curious features, and finally (they hope) coming up with a consistent theory to explain how Mercury was formed and how it came to appear as it does. A great deal of work has already been done and our knowledge of this hitherto little-known planet has been greatly advanced by the results from *Mariner 10*, but there is still a long way to go.

The Caloris Basin, the largest structural feature on Mercury discovered by Mariner 10, *with a diameter of 1300 kilometres, is on the right of the photograph. The ridged and cracked floor is bounded by a ring of mountains which indicates that the centre of the basin is just off the edge of the picture.*

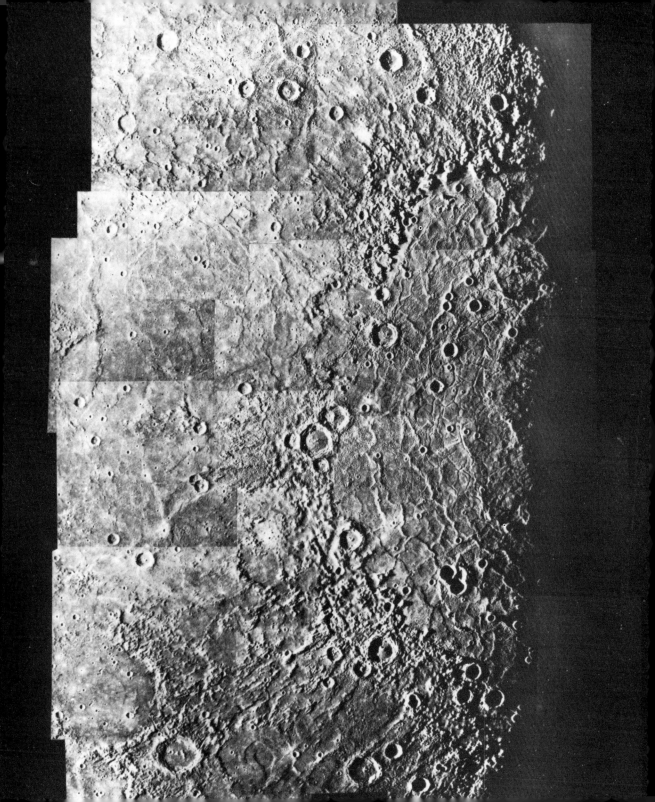

4 · Venus

Orbital period: 225 days; *rotation period*: 243 days; *distance from Sun*: 109×10^6 km (maximum) 108×10^6 km (minimum); *diameter*: 12,400 km; *mass*: 4.87×10^{24} kg; *density*: 4.9×10^3 kg m^{-3}; *satellites*: none.

The view from Earth

Despite being known as 'the evening (or morning) star', Venus is not a star but the second planet out from the Sun. Like Mercury, Venus is an 'inferior' planet as opposed to Mars, Jupiter, Saturn etc. which are 'superior' planets. This is not intended to be a comment on their characters but merely a way of distinguishing those that are inside the Earth's orbit from those that are outside. Since it is always closer to the Sun than the Earth, Venus can never be seen in the midnight sky but appears as a brilliant object in either the morning or evening sky, depending on which side of the Sun it happens to be. The diameter of the orbit is nearly twice that of Mercury, so Venus can appear in the sky nearly twice as far from the Sun as Mercury, up to a maximum angle of 48 degrees. Whereas Mercury is difficult to see, Venus is very hard to avoid since, after the Sun and Moon, it is the brightest object in the sky. It is seen at its brilliant best soon after sunset or just before dawn, but is so bright that it can be seen with the naked eye even in broad daylight, so long as you know exactly where to look.

It takes Venus 225 days to orbit the Sun, but 584 days to bring it back to the same position as viewed from the Earth. (These are the sidereal and synodic periods, explained in Chapter 3.) From eastern elongation to western elongation takes only 144 days during which time Venus passes between the Earth and Sun, but from western elongation back to eastern takes 440 days while Venus goes round the back of the Sun. This difference in time is purely a matter of the geometry of elongations and has nothing to do with the speed of Venus in its orbit (see diagram p. 37). In fact, Venus moves at an almost constant speed in the most nearly circular of all the planetary orbits.

Venus comes closest to the Earth at *inferior conjunction*, when it is between the Sun and the Earth. It approaches within 42 million kilometres which is very close indeed by astronomical

The changing appearance of Venus as seen from the Earth (see also the illustration on page 12).

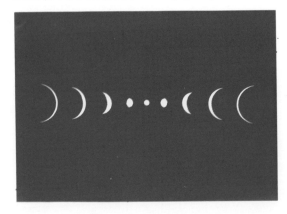

ABOVE *The Moon and Venus seen just after sunset; compare their brightness.*

standards; only the Moon and a few bits of interplanetary debris come any closer. At *superior conjunction*, when Venus is again on the Earth-Sun line but this time on the far side of the Sun, it is at a distance of over 250 million kilometres. Naturally, this vast change in distance affects the apparent brightness of Venus but not in as simple a way as one might expect. For example, maximum brightness occurs not when Venus is nearest to the Earth but 36 days before or after. This is because, like Mercury and the Moon, Venus exhibits phases. When closest to the Earth, Venus is 'new' and only the side away from the Earth is illuminated. At full, Venus is six times as far away, so not too bright. The maximum brightness occurs when Venus has the same

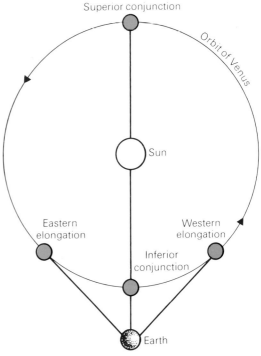

RIGHT *Seen from the Earth, Venus appears to be at its greatest distance from the Sun when it is at eastern or western elongation.*

crescent appearance as a five-day-old Moon. Some people with very good eyesight can detect the crescent phase with the naked eye, but it was Galileo who first studied the full range of phases with a telescope, and used this evidence to argue against the Ptolemaic theory (see Chapter 1).

When Venus passes between the Earth and the Sun it usually goes either above or below the exact Earth-Sun line. However, very occasionally the alignment is so close that Venus passes in front of the Sun and can be seen as a small, sharp spot on its surface. This is a *transit of Venus* and the last ones occurred in 1874 and 1882. The next ones will occur in 2004 and 2012, but if you miss those, you will have to wait until the year 2117 for the next transit. It is a bit like queueing for a bus!

The first observation of a transit of Venus was made by Jeremiah Horrocks, a young Lancashire clergyman who was also a brilliant astronomer. Horrocks calculated that a transit would occur on Sunday, 24 November 1639 at about 4 p.m. (not a good day for a clergyman), but the exact time was rather uncertain. He observed the Sun for most of Saturday, and then again from sunrise on Sunday except when 'called away . . . by business of the highest importance, which for these ornamental pursuits I could not with propriety neglect'. His piety was rewarded, for at 3.15 p.m. when he was again back at his telescope 'the clouds, as if by Divine interposition were entirely disposed then behold . . . a spot of unusual magnitude and of a perfectly regular shape which had already fully entered upon the Sun's disc'. Horrocks died suddenly two years later—in his twenty-second year.

Edmond Halley never saw a transit of Venus, but he recognized its importance as a method of measuring the size of the Solar System. To do this you need to measure just one interplanetary distance, as explained in Chapter 1, and then all the rest falls into place. Because Venus is much nearer to the Earth at times of transit than is the Sun, the apparent path of Venus across the Sun's disc will vary depending on where it is observed from on Earth. By comparing the appearance of a transit as viewed from two widely separated sites, and knowing how far apart the sites are, the distance of Venus can be calculated. Halley worked out exactly what was required but had to leave the actual experiment to his successors, as he would have been 104 years old when the next transit occurred in 1761. Expeditions to remote parts of the world were organized for both the 1761 and 1769 transits of Venus (including one led by Captain Cook) and, although several of them were frustrated by cloud, sufficient data were obtained for a good estimate of the Earth-Sun distance to be made.

Although Venus is well worth looking at with a telescope or binoculars, disappointingly little can be learned this way about the planet itself. (Once again, do not try to look at Venus when it is close to the Sun.) This is because the whole planet is always covered by a dense, featureless layer of cloud. One characteristic that can be measured is its diameter, which is about 5 per cent less than that of the Earth. Another is the *albedo*, or the proportion of light that is reflected. At 71 per cent this is the highest of any of the planets—a result of Venus's white, highly reflective clouds. It is this as much as the nearness of Venus that makes it such a bright object in the night sky.

Like Mercury, Venus has no natural satellites, so, until the Russians and Americans provided it with artificial ones, its mass had to be deduced from the small disturbances it made to the orbits of other planets. Again, it is not unlike the Earth, being about 82 per cent as massive. Similarly, the planet's density (obtained by dividing the mass by the volume) and surface gravity are close to those of the Earth—96 per cent and 90 per cent respectively. Altogether Venus appeared to be a slightly smaller version of the Earth, perhaps a little warmer and certainly more cloudy, but not at all a bad place for a holiday. It might also be (dare one suggest it?) a place where intelligent life had evolved. We could forgive the Venusians for

Venus and Mars painted by Botticelli.

not having tried to make contact because, with their constantly cloudy skies, they would be unaware of the existence of other planets. Just how far this idyllic picture was from the truth was only fully revealed when spacecraft were sent to Venus.

Spacecraft exploration

As the nearest of the planets, Venus is an obvious choice for exploration by spacecraft, and is second only to the Moon in the number of visits it has received from Earth. The first was as early as December 1962 when NASA's *Mariner 2* flew by at an altitude of 35,000 kilometres—about the distance of a weather satellite from Earth. The first really close encounter came in 1966, when the Russian spacecraft *Venera 3* actually reached the planet's surface, but unfortunately it failed to send back any data. Since then, the *Venera* series of spacecraft has been hugely successful, from *Venera 4* which transmitted data for 94 minutes while it parachuted to within 40 kilometres of the Venusian surface, through *Veneras 5, 6, 7* and *8,* which all continued to function after landing (for periods which increased up to 50 minutes as the design was modified in the light of data sent back from earlier missions), *Veneras 9* and *10* which made soft landings and sent back photographs of the surface, right up to *Veneras 15* and *16* which were placed in orbit around the planet in 1983 to carry out radar surveys of the surface.

To appreciate just what an achievement all this has been, one has to realize that Venus is not such a welcoming planet as was at one time thought. The soft clouds are not composed of water vapour, but of sulphuric acid and chlorine compounds. There is a large amount of carbon dioxide in the atmosphere which lets the Sun's heat in more readily than it lets it escape (rather like the glass in a greenhouse), so the surface is very hot indeed—hot enough to melt lead. At the surface the atmospheric pressure is nearly a hundred times as great as on the Earth. Just imagine the problems of designing a delicate piece of apparatus that can drop out of a vacuum, through an acid bath and into an atmosphere exerting the same pressure as is experienced at the bottom of the deep ocean, to crash amongst searing hot rocks and still transmit data over 40 million kilometres back to Earth.

The early *Veneras* were concerned with measuring the temperature, pressure and composition of the atmosphere, and how these varied with height. When it was found that the cloud was not so thick that it prevented light reaching the surface and that the clouds did not extend all the way down to ground level, it was realized that it might be possible to obtain pictures (providing a camera could be designed to withstand the landing, and did not mind an oven-hot lens). This was achieved in 1975 when both *Venera 9* and *Venera 10* sent back good quality pictures showing a rocky terrain extending away to the horizon.

While this was going on, the Americans had had three missions to Venus: *Mariner 2* (already mentioned), *Mariner 5* and *Mariner 10* (which swung by Venus on its way to Mercury, see Chapter 3). These had sent back fine ultra-violet pictures of the tops of the clouds, measured their temperature and established that Venus had a negligible magnetic field. While this was no mean achievement, it was hardly in the same league as the *Venera* programme. However, the next NASA mission—*Pioneer Venus*—certainly was. On 20 May 1978, the first part of *Pioneer Venus*, a spacecraft designed to go into orbit around the planet, was launched from Cape Canaveral. Three months later the second part, the *Multiprobe*, was sent off in pursuit. Both parts arrived at Venus on 9 December, when the orbiter duly went into orbit and the *Multiprobe* divided into five separate parts, each destined to land on a different area of Venus. The probes made their measurements on the way down and were not intended to survive the impact (though, surprisingly, one of them did, and continued to operate for over an hour). The orbiter had a much longer life, studying the upper atmosphere from increasingly close range as its orbit was adjusted until it was actually passing through the upper atmosphere at a height of 150 kilometres.

Since then two Russian *Vegas*, on their way to encounter Halley's Comet, dropped off what were effectively meteorological balloons to sink slowly down through the Venusian atmosphere, measuring winds and other meteorological phenomena.

As a result of all this intensive activity, supplemented by radar studies from Earth, we now have a very much clearer idea of what Venus is like. The rotation period is very close indeed to two-thirds, not of a Venusian year, but of an Earth year. There are too many such numerical relationships in the Solar System for them to be mere coincidence, so it seems that it is the Earth which controls the length of the Venusian day, possibly through some tidal interaction. Another curious feature of the Venusian rotation is that it is the opposite way to that of the other planets, so the Sun (if it could be seen through the dense clouds) would appear to rise in the west and set in the east.

In spite of the solid cloud, which occurs at three distinct levels, it has been possible to map the surface in some detail using radar. Obviously,

The cloud-covered face of Venus as seen by Mariner 10 *on its way to Mercury.*

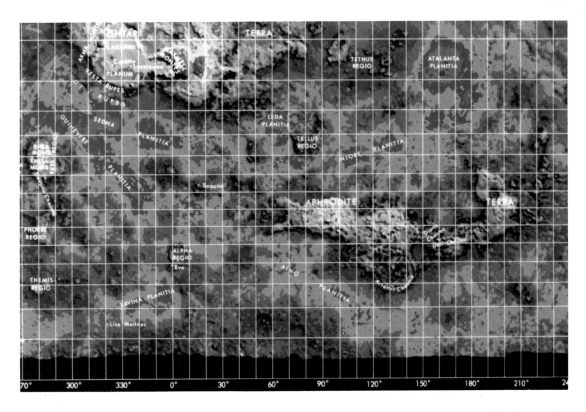

A relief map of the surface of Venus, giving the names of some of the major features.

there are no oceans because of the temperature (there is not a great deal of water, anyway, even in the atmosphere) but there are mountains. The biggest is named after Maxwell, the great physicist, but he is the only male represented there, as nearly all the other features, appropriately enough, are given feminine names. These are mostly mythological such as Diana, Rhea, Lakshmi and Kesta, but a few real women are included, such as Cleopatra, Colette and Lise Meitner. The features include craters, canyons, cliffs, plateaus and two regions of higher ground known as continents (Aphrodite and Ishtar). Although sufficient sunlight filters through the cloud for it to be light by day and dark at night, there is very little temperature variation during the day, and not much in the way of winds near the surface. The weather is not completely bland, though, as lightning has been detected.

We can no longer think of Venus as a sort of home-from-home, but now know it to be 'very much like a classic view of hell', as one space scientist described it. It may also provide a timely warning. Most of the unpleasant conditions on the surface result from the searing heat, and this is a direct result of the carbon dioxide in the atmosphere which causes a 'runaway greenhouse effect'. Perhaps we should be more careful about the way we pour carbon dioxide into our own atmosphere to avoid a similar fate overtaking the Earth.

5 · Earth

Orbital period: 365 days; *rotation period*: 23.93 hours; *distance from Sun*: 152×10^6 km (maximum), 147×10^6 km (minimum); *diameter*: 12,742 km; *mass*: 5.97×10^{24} kg; *density*: 5.5×10^3 kg m^{-3}.
Satellite Moon; *orbital period*: 27.32 days; *radius of orbit*: 384×10^3 km; *diameter*: 3476 km; *mass*: 7.34×10^{22} kg; *density*: 3.3×10^3 kg m^{-3}.

By far the most interesting thing about the Earth is that it is the only place in the Universe known to support intelligent life. Even without this strange property, the Earth would still be a fascinating planet (fascinating to whom you might ask!), but let us start with some of the more basic facts.

It is the $23\frac{1}{2}°$ tilt of the Earth's rotation axis that gives us the seasons. In December the northern hemisphere is inclined away from the Sun and experiences long nights and short days. In June the position is reversed and it is summer in the northern hemisphere. At the equinoxes the Sun is above the Equator and day and night are the same length.

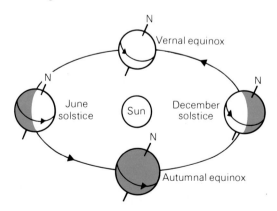

The Earth is the third planet out from the Sun. Its orbit is not quite circular so the distance from the Earth to the Sun varies. It is 5 million kilometres less in January than in July. Why, then, is it colder in January than July, at least in the northern hemisphere? This is because the Earth's axis of rotation is not at right-angles to the plane of the orbit, but differs from a right-angle by $23\frac{1}{2}$ degrees. As the Earth goes round the Sun, the axis continues to point towards the same point in space (close to the direction of the Pole Star), so that in June the northern part of the Earth is inclined towards the Sun and in December it is inclined away. Thus, for people living in Greenwich in London the Sun rises high in the sky in June and we receive its rays nearly head-on, whereas in December it never gets far above the horizon and barely warms us at all. For the southern hemisphere it is the other way round, so the seasons are reversed. In March and September the Sun is over the Equator, favouring neither north nor south, and we have the equinoxes, when day and night are equal in length.

The size of the Earth was first measured by—you've guessed it—a Greek astronomer, but he did it in Egypt. In about 250 BC Eratosthenes noted that the midsummer sun was so nearly overhead in Aswan (then called Syene) that it shone directly down a well, whereas at Alexandria, some 800 kilometres further north, the Sun came no closer than about 7° to the vertical. He correctly deduced that this is because the Earth is spherical and, making the (correct) assumption that the Sun is at a huge distance compared with the size of the Earth, he reasoned that if 7° is a fiftieth of a circle, then 800 kilometres is a fiftieth of the Earth's circumference. Of course, he worked in ancient units (stades) rather than kilometres but he got an answer which is remarkably close to the truth.

The metre was not introduced until 1791, when it was defined in France as one ten-millionth of the distance from the North Pole to the Equator. The circumference of the Earth was now accurately known in metres (40,000 kilometres), but this did not help much until someone had measured it up to find out what a metre actually was. The Earth is not a perfect sphere, but bulges slightly at the Equator and is flattened at the poles. The distance between the poles is about 43 kilometres less than the diameter of the Equator.

Since the Earth has a moon, the mass of the Earth can be determined by putting the distance of the Moon ($d = 3.8 \times 10^8$ metres) and the length of time it takes to complete an orbit ($t = 2.4 \times 10^6$ seconds; one lunar month) into the formula $M = 5.9 \times 10^{11} \times d^3/t^2$, where M is the mass in kilogrammes. This method, which is a direct consequence of Newton's law of gravity, can be used to determine the mass of any large body orbited by a small one—for example the Sun's mass from the length of the year (in seconds) and the distance of the Sun (in metres)—try it. Since we have the advantage of living on the Earth, there are several other methods available to us for measuring its mass, but the one just described is perfectly good.

The Earth's average density is the highest of all the planets. It is also much higher than the average density of surface rocks, so there must be something pretty dense deeper down. Studies of earthquakes have revealed that the Earth is reasonably uniform from just below the surface to nearly half-way to the centre. This part is called the *mantle*, and it is about as rigid as concrete. Below this there is a sudden change from solid to liquid when the Earth's core is reached. Although the core is over half the diameter of the Earth, it is only about one sixth of its volume and is very dense indeed. It is probably composed mainly of molten iron. Within 800 kilometres of the centre the core becomes solid again.

As the Earth's mantle is as rigid as concrete, we do not need to worry about sinking into it. At least, not in the short term. However, given a time-span of millions of years, even the mantle material can manage to flow, and it is believed to be convecting very slowly (i.e. the warm parts are moving up and the cooler parts moving down, like water being heated in a pan). When an upward-moving region reaches the surface, it has no choice but to spread out, and similarly a downward-moving flow sucks in nearby surface material. The part of the Earth we actually live on is a thin layer of scum a few tens of kilometres thick called the *crust*, which is floating on and getting dragged around by the mantle. The crust consists of a dozen or so fairly rigid plates that are constantly moving relative to one another, sometimes separating (e.g. those underlying North America and Europe), sometimes colliding head-on (e.g. India and Asia), sometimes just sliding past each other. This movement, known as *plate tectonics*, is responsible for many of the large-scale surface features of the Earth: the Himalayas have been thrown up at the site of a collision, while the Atlantic is a region where the mantle material is spreading out, leaving a line of volcanic features down the middle, the most obvious of which is Iceland. The famous San Andreas fault through San Francisco results from two plates sliding past each other; similarly, the deep-seated earthquakes around Japan are associated with a plate on the Pacific side trying to slide under the Asian plate, and so on.

Red-hot volcanic lava engulfing trees on a snow-covered slope of Mount Etna.

Nearly all earthquakes occur at plate boundaries where adjacent plates are rubbing against one another, and volcanoes also tend to be associated with the edges of plates. Britain is fortunate to be well in from the edge of its plate, so it has little to fear from volcanoes or earthquakes. Many people think that the lava that pours out of volcanoes comes from the molten core, but this is not so. The lava comes from within a few kilometres of the surface, and is less dense than most surface rocks. It is normally solid, because it is under great pressure from the overlying rocks, but when there is a crack in the crust the pressure is released, enabling the volcanic material to melt and squeeze up to the surface. Once it emerges—often spectacularly—it cools and solidifies.)

It is not only the effects of plate tectonics that shape the surface of the Earth. Less spectacular, but equally important, is the process of weathering. Rocks at the surface are constantly bombarded by wind, rain and waves, or affected by the action of frost, rivers, moving glaciers and the Sun's radiation. Each of these in its own way tends to erode away the surface, breaking it up into small pieces that are then transported by wind, rain and rivers from higher to lower ground. Much of the material ends up in the sea, where it gradually settles and compacts until further tectonic activity raises it up again. Early geologists found it difficult to believe in the uplift of material, but how else could they explain the presence of fossilized sea-creatures in rocks, even in the highest mountains? The process of uplift, weathering, transportation and deposition is known as the geological cycle and this is what has produced the appearance of the Earth as we see it today. The process is still going on, but on a time-scale so much longer than a human life that we barely notice it.

We now come to the feature that singles out our planet from all the others—it is covered in water. Well, not quite covered, but a good two-thirds of the surface is ocean. The Pacific Ocean alone covers nearly half the Earth, as you can see if you look at a globe. Similarly, the view from above the South Pole shows a great deal of ocean and very little land: most of the land is concentrated in one half of the northern hemisphere. The oceans are very much more than a reservoir for rain-water, though that is one of their useful functions. Their ability to absorb and re-radiate heat from the Sun is a major factor in determining the temperature of the Earth and in preventing huge variations in temperature from day to night. It is also likely that life first emerged in the oceans. In all these processes, and in the production of weather, there is a close connection between the oceans and the atmosphere, with the power being supplied by the Sun.

The Earth's atmosphere is pretty thin compared to that on Venus, Saturn or Jupiter, but much thicker than on Mars or Mercury. As a useful comparison, the weight of the atmosphere overhead and the pressure it exerts are just the same as would be experienced if it were water 9 metres deep. Put another way, you have about 300 kilogrammes of air resting on your head! Until the advent of rockets and satellites, all we could ever see of the stars and planets was the light that managed to penetrate this rather substantial barrier. Fortunately, it is transparent to the light to which our eyes are sensitive, but it filters out nearly all the other wavelengths, such as ultra-violet, infra-red and most of the higher frequency radiation that would be harmful to us. Perhaps 'fortunately' is the wrong word, because naturally enough our eyes have developed to cope with the light that *does* reach them.

As astronomers are only too well aware, the atmosphere carries quite a lot of water, either as water droplets (clouds and rain) or as water vapour. The amount of water vapour in the air varies a good deal, but it can be as high as 2 per cent. The remainder of the atmosphere consists of 79 per cent (by volume) nitrogen, 20 per cent oxygen, 1 per cent of the inert gas argon and a trace of carbon dioxide. The remaining constituents are measured in parts per million. The minor components can have a significance out of all proportion to this quantity: we have already seen how carbon dioxide can produce a runaway greenhouse effect on Venus, and to a very much smaller extent it does the same on Earth, reducing the amount of heat that escapes back into space.

As one goes higher, the atmosphere gets thinner, which is why observatories are usually

The planet Earth viewed from space.

Even the most ordinary picture can show what makes the Earth such an extraordinary planet: a solid surface with large areas of open water on it; an atmosphere containing water-vapour clouds; plants, animals and artificial structures.

sited at the tops of mountains; at the top of Everest there is more atmosphere below you than above. But even at great altitudes there is still a lot going on. The high-energy radiation from the Sun penetrates to within about 100 kilometres of the surface of the Earth and on its way down it breaks up the atmospheric particles into electrons and positively charged nuclei. These electrically charged particles form the *ionosphere*. Beyond the ionosphere is the *magnetosphere*, where the movements of the few remaining atmospheric ions are controlled by the magnetic field. This extends out to 60,000 kilometres from the Earth on the sunward side, where it is compressed by the solar wind, and much further on the night side. It is only beyond that that one has truly left the Earth's environment and entered interplanetary space.

Earthrise seen from the Moon.

6 · Mars

Orbital period: 687 days; *rotation period*: 24.62 hours; *distance from Sun*: 249×10^6 km (maximum), 207×10^6 km (minimum); *diameter*: 6870 km; *mass*: 6.42×10^{23} kg; *density*: 4.0×10^3 kg m^{-3}.

Satellites

Name	Orbital period (hr)	Radius of orbit (10^3 km)	Dimensions (km)	Mass (10^{15} kg)	Density (10^3 kg m^{-3})
Phobos	7.654	9.378	$27 \times 22 \times 19$	9.6	1.9
Deimos	30.030	23.459	$15 \times 12 \times 11$	2.0	2.1

The view from Earth

The fourth planet out from the Sun is Mars, the 'red planet'. Since its orbit is greater than that of the Earth, it is a superior planet and makes a complete circuit of the sky instead of always staying close to the Sun. As seen from the Earth, it completes a circuit in 780 days, or just over two years. Most of the time it travels from west to east against the star background, but for 70 days it reverses its direction and goes from east to west. This *retrograde motion* occurs when Mars is nearest to the Earth and when the greater orbital speed of the Earth allows it to overtake Mars. All the superior planets show retrograde motion as the Earth passes between them and the Sun.

When Mars is nearest to the Earth it is brighter than any star and is very easy to recognize because of its reddish hue. Its closest approach occurs at *opposition*, when the Earth is between the Sun and Mars and all three are in line. At this time Mars will be south at midnight. As Mars proceeds in its orbit it goes through *quadrature*, when it is at right-angles to the Earth–Sun line and is south at 6 a.m., and then on to *conjunction*, when it is on the opposite side of the Sun to the Earth. The word that covers both opposition and conjunction is *syzygy*, but this is seldom used except by Scrabble players. Mars continues on its way, passing through eastern quadrature before performing its little retrograde dance round about opposition.

Clearly, the best time to observe Mars is at opposition, when it is closest and when the Sun is safely set. But some oppositions are better than others, because the orbit of Mars is elliptical and its distance from the Sun varies between 200 and 250 million kilometres. Subtracting the Earth–Sun distance shows that Mars at opposition can be anything between 50 and 100 million kilometres away from the Earth.

At its nearest, Mars appears as large as a 1p coin at a distance of 150 metres—too small to be more than a point to the naked eye, but easily seen as a disc through binoculars or a telescope. The first thing to notice is that it is much redder than when seen by eye. This is because the extra light collected by the telescope is enough to actuate the colour-sensitive cones in the eye (see Chapter 2). Also it should be possible to see the whiter patches at the top and bottom. These patches are near the poles of Mars and change their size throughout the martian year. At first they were thought to be ice-caps, like those on the Earth, but later they were thought to be thin coverings of frost, composed of either water or carbon dioxide. Finally, it might just be possible to see darker markings on the surface. Maps of the planet were made by studying these markings carefully over many nights, and their change in position with time was used to calculate the rotation period of Mars, which was found to be very similar to that of the Earth.

There are slight changes in the colour and

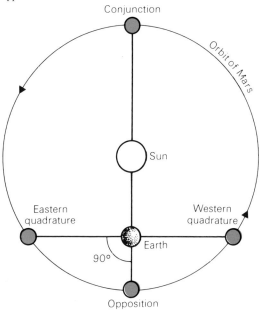

BELOW *The relative positions of the Sun, Earth and Mars as Mars goes from opposition through western quadrature, conjunction, eastern quadrature and back to opposition.*

ABOVE *The retrograde motion of Mars. Near opposition, Mars moves from east to west against the stars as a result of the greater orbital speed of the Earth.*

LEFT *Mars and Jupiter over the Old Royal Observatory, Greenwich. Mars is the fainter planet, to the right of Jupiter.*

markings of Mars which led Schiaparelli, after very careful observation, to announce in 1877 that he could see a series of straight lines which he called *canali* (Italian for channels). This suggested what the public was anxious to believe, that there was life on Mars. The belief was fed by science-fiction writers, who visualized a race of beings on a cold and dying planet carefully conserving the last arable regions from the encroaching desert by means of elaborate irrigation canals which made the best use of the meagre water supply. Even the failure of some astronomers to see the canals was used to reinforce this point of view; it was not the canals themselves that were seen—they would be far too narrow to show up—but the vegetation surrounding them. Clearly this would vary with season and could not always be seen. Also it should be remembered that such fine details could be seen only at favourable oppositions, so for most of the time the theories could flourish unhindered by observation.

Had Schiaparelli been the only man to see canals, or had he given them a less suggestive name, they might have been quietly forgotten, but they were also seen by other eminent astronomers. One of these was Percival Lowell, whose book *Mars and its Canals* (published in 1905) states: 'that Mars is inhabited by beings of some sort or other we may consider as certain.' Other astronomers were more sceptical, both about the canals and the martians. The canal controversy was not finally resolved until direct evidence was available from spacecraft.

Mars has two moons, though they are so small that both of them would easily fit into one crater of our Moon. They were discovered in August 1877 (during the same close approach when Schiaparelli mapped his canals) by the American astronomer Asaph Hall, who named them

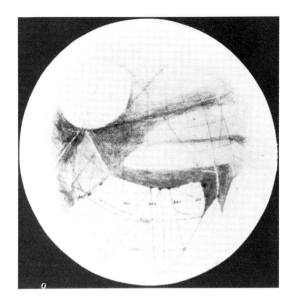

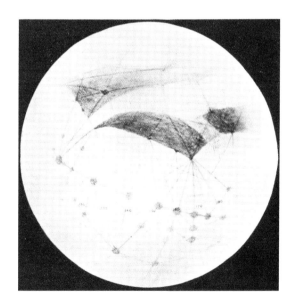

The canals of Mars as depicted by Percival Lowell.

THE PLANETS

Mars's smaller satellite, Deimos, which is only 15 kilometres from end to end.

Phobos and Deimos after the mythological attendants of Mars. Both satellites are in nearly circular orbits close to Mars's equator and both orbit in the same direction as Mars spins. As Deimos, the outer satellite, has an orbital period only slightly larger than Mars's period of rotation, it almost manages to keep up with Mars and to an observer on the planet would appear to be in more or less the same place throughout the night. In that time it would change its phase and very likely be eclipsed, but none of this would be very spectacular as it would appear to be only about one-fourteenth of the diameter of our Moon (about the size of a 1p coin at a distance of 6 metres) and the whole of its surface would give out less light than does Venus on Earth.

Phobos would be a more interesting object for an observer on Mars. It is larger than Deimos (though still only 27 kilometres from end to end) and a lot closer to the planet, so close in fact that it could not be seen by an observer more than 70° from the Martian equator, as it would always be below his horizon. If it was viewed from the equator when directly overhead, it would appear to be about a third of the diameter of our Moon and an observer would probably be able to see craters and lines on its surface (would he believe them to be canals?). Because Phobos orbits faster than Mars spins, it would rise in the west and set in the east, and would pass from western horizon to eastern horizon in little more than 4 hours, going through more than half its cycle of phases in that time. More likely than not it would pass through Mars's shadow and be eclipsed. It also frequently eclipses the Sun, but is only large enough to cut off a quarter of the light. It does this 1300 times a year, but each solar eclipse lasts no more than 19 seconds.

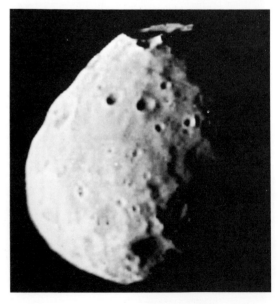

Mars's larger and nearer satellite, Phobos.

From the orbital period and distances of Phobos and Deimos we find that the mass of Mars is only 11 per cent of that of the Earth, though its volume is 16 per cent. Thus its average density is less than that of our planet and this suggests that the central core of Mars is much smaller than the Earth's core.

Spacecraft exploration

As with most of the planets, our knowledge of Mars has increased immeasurably since the advent of spacecraft, but the information was not easily won. Considering Mars is called the red planet, it has been very unfriendly to the Russians. Only two of their nine Martian spaceprobes have achieved any degree of success, sending back data on the temperature, clouds and topography. The American programme has been more successful and all but one of the five *Mariner* spacecraft they sent to Mars obtained useful data. The first three showed that there was almost no magnetic field and an atmospheric pressure of less than one per cent of that on Earth. They also obtained television pictures of 20 per cent of the Martian surface showing plenty of craters, but no canals. By sheer bad luck, this happened to be just about the dullest 20 per cent and it gave what we now know to be the false impression that Mars is a very dull planet indeed.

The fifth American spacecraft was *Mariner* 9 which sent back over 7000 high quality pictures during its working life in orbit from November 1971 to October 1972. These covered just about all the Martian surface. When *Mariner* 9 started work there was a huge dust storm on Mars and this obscured much of the surface, but the atmosphere eventually cleared to reveal all the Martian wonders. It is true that there was no sign of life, but there were huge mountains, one of which (Olympus Mons) is a giant volcano rising to 23 kilometres (three times the height of Everest) and over 500 kilometres in diameter, bounded by cliffs 6 kilometres high. It is similar to the 'shield' volcanoes on Hawaii but much, much larger.

There were other volcanoes, a large number of impact craters, gently rolling plains between the cratered areas which might well be ancient lava-flows, saucer-like depressions, vast canyons and, most surprising of all, long sinuous channels that could only have been formed by water-flow. Close inspection shows that, as well as the expected evidence of wind erosion, there is also evidence of a great deal of water erosion, indicating that Mars was at some time a lot wetter than it is now.

So what has happened to all the water? This brings us back to the polar caps, which are now believed to be substantial layers of ice, perhaps a kilometre thick. The old theory about them being very thin layers of carbon dioxide frost is not dead though as such a layer probably does form during the Martian winter, and it is this which is seen to expand and contract with the seasons, while the ice layer remains fairly constant.

After *Mariner 9*, there was not much more that could be found out about the surface of Mars without actually landing on it. This was achieved by the American *Vikings 1* and *2* in July and September 1975. The design of the *Vikings* was based on that of the Mariners but, in addition, each carried a module that could descend from orbit and make a soft landing. Although the expression is an insult to the vastly sophisticated systems involved, everything went like clockwork. The spacecraft went into orbit around Mars and spent some days searching for suitable landing sites. Each landing module then separated from its orbiter and was set down with no more violence than you would experience when jumping off a chair.

Though widely separated, the view from both landers was very similar—a rust-coloured, boulder-strewn desert stretching away to the horizon beneath a pink sky. There was no sign of any life, but a random landing on Earth would be unlikely to show living things. There were more subtle life-detectors on board though including

The giant volcano Olympus Mons rising majestically above the Martian cloud (artificial colour).

one that could detect minute quantities of organic material, alive or dead, that might have blown to the landing area from any part of the planet. It found nothing. In addition there were three more independent life-detecting experiments which involved taking soil samples, giving them warmth and nourishment to encourage life processes, and then analysing the results. The results of these experiments were less clear-cut and not so easy to interpret as had been expected, so the question of life on Mars still remains open, although the bulk of the evidence suggests that none exists, even of the most primitive kind.

The landers also carried other equipment, including a meteorological station that detected nothing more violent than gentle breezes, though the presence of sand dunes, the erosion of the nearby rocks and the dust storm that drifted over from the other hemisphere showed that conditions were not always calm. The temperature varied from a bitterly cold $-32°C$ in mid-afternoon down to $-113°C$ just before dawn. The atmospheric pressure was confirmed as being around 7 millibars (in comparison with 1000 millibars on Earth), but was found to vary quite iron. The characteristic reddish-brown colour of the planet is due to a thin coating of iron oxide on the rocks—Mars is rusty! The pink sky results from the fine particles of dust left suspended in the atmosphere after the dust storms.

While the landers were conducting their experiments, the Viking orbiters had not been idle. Besides acting as relay-stations for the lander data, they had been busy mapping the surface of Mars at a higher resolution than *Mariner 9* and sending back magnificent pictures, notably one of Olympus Mons wreathed in cloud. On the way in they had also taken close-up pictures of the Martian satellites (as had *Mariner 9*), showing them to be irregular, black and elongated and pocked with craters. The pictures also showed them to be Mars-synchronous: like our Moon, they always have the same side facing the planet.

ABOVE *Mars from* Viking Orbiter 2, *August 1976. At the top, with water-ice cloud plumes on its western flank, is Ascraeus Mons, one of the giant Martian volcanoes. In the middle is the great rift canyon called Valles Marineris and near the bottom is the large, frosty crater basin called Argyre.*

BELOW *The barren, rocky surface of Mars as seen just before sunset by* Viking Lander 1 *in 1976. The larger nearby rocks are about 30 centimetres across. Note the orange sky.*

strongly with season, presumably as a significant proportion of the atmosphere is deposited as frost at certain times of the year. Analysis of the atmosphere showed it to be, like that of Venus, nearly all carbon dioxide, though with a rather different selection of minor constituents.

Chemical analysis of the soil showed it to be not dissimilar to that found on Earth, though rich in

7 · Jupiter

Orbital period: 11.86 years; *rotation period*: 9.92 hours; *distance from Sun*: 816×10^6 km (maximum) 741×10^6 km (minimum); *diameter*: 143×10^3 km (equatorial), 134×10^3 km (polar); *mass*: 1.90×10^{27} kg; *density*: 1.3×10^3 kg m^{-3}.

Satellites

	Name	Orbital period	Radius of orbit (10^3 km)	Diameter (km)	Mass (10^{18} kg)
XVI†	Metis	7.07 hr	128	40	0.1
XV	Adrastea	7.16 hr	129	25 × 20 × 15	0.02
V	Amalthea	11.95 hr	181	270 × 166 × 150	7.2
XIV	Thebe	16.19 hr	222	110 × 90	0.8
I	Io	1.769 d.	422	3,630	88,900
II	Europa	3.551 d.	671	3,138	47,900
III	Ganymede	7.155 d.	1,070	5,262	148,000
IV	Callisto	16.69 d.	1,883	4,800	107,000
XIII	Leda	238.7 d.	11,094	16	0.006
VI	Himalia	250.6 d.	11,480	186	9.4
X	Lysithea	259.2 d.	11,720	36	0.08
VII	Elara	259.7 d.	11,737	76	0.8
XII	Ananke	1.73 yr*	21,200	30	0.04
XI	Carme	1.89 yr*	22,600	40	0.1
VIII	Pasiphae	2.01 yr*	23,500	50	0.2
IX	Sinope	2.08 yr*	23,700	36	0.08

*Retrograde orbit.
†The Roman numerals are assigned to the satellites in the order of their discovery.

The view from Earth

Although second in brightness to Venus, Jupiter is very easy to see as it shines prominently in the night sky for most of each year. Once it has been located, it is easy to find again as it hardly moves from night to night, making a stately progress through the zodiacal constellations at the rate of one per year, until it completes the full cycle in a little under 12 years. Jupiter is the next planet out after Mars (disregarding the asteroids—see Chapter 10), but it is over three times more distant from the Sun. The fact that it still appears so bright even at this distance suggests that it is very large, and this can easily be confirmed by looking at it through binoculars, when it will show an obvious disc. It appears as large as a 1p coin at a distance of 22 metres when it is at opposition diminishing to a 1p coin at 32 metres at conjunction, when it is on the far side of the Sun from the Earth. Taking account of its distance this means that Jupiter's diameter is eleven times greater than that of the Earth and that its volume is more than 1400 times greater. In fact, Jupiter is easily the largest object in the Solar System after the Sun, greater than all the other planets put together.

Galileo was the first man to look at Jupiter through a telescope. As well as seeing its disc, he also noticed that it was attended by no less than four moons that orbit around it in periods of between 2 and 17 days. Through binoculars the Galilean satellites just look like stars strung out on either side of Jupiter (because we are looking at their orbits edge-on), but a series of observations will show that they are not stars as they oscillate from side to side of the planet rather than standing still while Jupiter passes by. Besides their historical importance in showing that not everything revolves around the Earth, they also provided the first means of measuring the velocity of light and were an important aid to navigation.

The four Galilean satellites are quite important members of the Solar System in their own right; Callisto and Ganymede are as big as Mercury while Io is bigger than our Moon and Europa is only a little smaller. Jupiter has a further twelve satellites, four closer to the planet than the Galilean satellites and the others further out, but they are all pretty small. The twelve innermost satellites are well behaved and go round Jupiter in the same direction that the planets go round the Sun, but the four outer ones are *retrograde*, that is, they go round in the wrong direction. We will have more to say about Jupiter's satellites when we come to spacecraft exploration, but for the moment let us return to the planet itself.

Although Jupiter's mass is over 300 times that of the Earth, its density is only a quarter as great. If it were made of the same materials as the Earth, these would be even more compacted by Jupiter's huge gravity and the density would be greater. Like the Sun, it is composed mainly of hydrogen, with a little helium and small amounts of other elements. Unlike the Sun, it is not blazing hot, so the hydrogen is in the form of a gas only in the outer 1000 kilometres. Below that it forms a sea of liquid hydrogen and deeper still, perhaps halfway to the centre of the planet, the pressure is so great that the hydrogen changes into what is known as metallic hydrogen. This is also a liquid, but it has the additional property of being a good conductor of electricity. Jupiter may have a small, solid core at its centre.

With so much liquid hydrogen involved, you might think that Jupiter is extremely cold, and you would be right, at least as far as the

Jupiter through a small telescope, showing belts of clouds and two of the four Galilean satellites.

outermost layer is concerned. But Jupiter is known to radiate more heat into space than it receives from the Sun, so it must be somewhat warmer inside. In fact, the temperature of the interior of the planet is believed to go up to many thousands of degrees Celsius and the hydrogen manages to remain liquid only because of the enormous pressure. Since hydrogen and helium are not radioactive, Jupiter is not generating heat now. Its high temperature is a relic of its early history when the newly-formed planet was extremely hot and probably shone out like a mini-Sun. We might well ask why Jupiter did not turn into a star, since it has the same composition as the Sun. The reason is that it is not big enough to generate the temperature and pressure required to start a thermonuclear reaction. Jupiter may be thought of as a 'failed star', but if it had been a hundred times larger . . . .

Even through a small telescope it is easy to see that the disc of Jupiter is broken up into a series of bands running parallel to the equator. These are not surface features but cloud markings at the top of a very thick atmosphere. Jupiter also has a huge 'red spot'. This shifts around a bit, mostly in longitude, and sometimes shows up more obviously than at other times, but it has been visible for at least as long as telescopes have been good enough to show it. For a long time theorists attempted to interpret the red spot in terms of something on the surface, such as the outpourings from a volcano, or a column of turbulence associated with a mountain, but the present view is that it is simply a very long-lived cyclone. It may seem excessive for a cyclone to last for over 300 years, but it must be remembered that the atmosphere of Jupiter is very different from our own. It is believed to be 1000 kilometres deep, mainly composed of hydrogen, with crystals of ammonia, water and ammonium hydrosulphide as well as water droplets. Also it rotates very rapidly. Despite Jupiter's enormous size it spins around in less than 10 hours so that its equator moves about 90 times as fast as the Earth's. There

Voyager *picture of the great red spot, a giant cyclone in Jupiter's southern atmosphere with turbulent cloud downstream from it. The white area immediately below the great red spot is a similar, anticlockwise rotating storm.*

is still much to learn about Jupiter's atmosphere, in particular what causes the red and brown colours in it, but we can be sure that there are huge winds and a great deal of turbulence.

Spacecraft exploration

The voyage to Jupiter is ten times longer than a journey to Mars or Venus, so it is hardly surprising that fewer spacecraft have been there, but the four that have made fly-bys have sent back spectacular results. The first two were numbers 10 and 11 of the well-tried *Pioneer* series. *Pioneer 10* successfully flew past Jupiter in December 1973 at a distance of 130,000 kilometres and then continued its outward journey to become the first man-made object to leave the Solar System. It carries a plaque indicating where and when it was made and showing the strange creatures who built it, just in case it should be picked up by intelligent beings in the remote future in some other part of our Galaxy. *Pioneer 11* was targeted still closer to Jupiter—34,000 kilometres—and swung by in December 1974 to emerge on course for an encounter with Saturn five years later.

The *Pioneer* spacecraft were deliberately designed to spin in order to provide a suitable platform for many of the experiments, but a spinning platform is far from ideal for photography. Although the pictures taken from them were much better than could be obtained from Earth, they are nowhere near the quality of those obtained by the later *Voyager* spacecraft. The main purpose of the *Pioneers* was to assess the environment through which they travelled, first the asteroid belt (see Chapter 10), to see if it

contained any hazards for future missions, then the environment of Jupiter (its magnetic field, radiation, particles, temperature, etc.) and finally the environment beyond Jupiter. In addition to this the *Pioneers* detected helium in Jupiter's atmosphere and showed that Io has an atmosphere of its own, as well as obtaining data that refined our ideas of the planet's internal and atmospheric structure.

Some three years after the *Pioneer* fly-bys, two more spacecraft were heading for Jupiter. These were *Voyagers 2* and *1*, launched in that order in 1977, though *Voyager 1* overtook its twin to reach Jupiter first in 1979. Although these spacecraft were identical, they had differing and complementary missions. The Galilean satellites were divided between them (one spacecraft could not take in all four in a single fly-by), and they went on to obtain differing views of Saturn and its satellites in 1980 and 1981. *Voyager 1* was then destined to head off into space while *Voyager 2* visited Uranus and Neptune. Despite numerous

THE PLANETS

Voyager, *one of two identical spacecraft launched in 1977 to investigate the outer planets. The spacecraft weighs about 800 kilogrammes and includes a dish antenna for communicating with Earth, a nuclear generator for providing power (on small boom, lower left), cameras and other instruments on the right-hand boom and a magnetometer at the end of the long boom.*

minor problems both before and after launch, the two *Voyagers* duly arrived at Jupiter in March and July 1979. They sent back a stunning series of pictures of Jupiter's atmosphere which have been combined to form a film showing a speeded-up version of the wild swirling movements, particularly in and around the great red spot. The spot itself was found to rotate in about six days, and to be cooler than its surroundings. Io was found to be linked to Jupiter by a magnetic field along which flowed a current of more than a million amps, giving rise to strong radio-wave emissions. Aurorae and lightning were seen. Temperatures were measured in the atmosphere, which was found to be 11 per cent helium.

If the planet was yielding up its mysteries, observation of its satellites produced a whole range of new ones to puzzle the scientists. Callisto is about the same size as Mercury and is just as heavily cratered, but there the similarity ends. It is much less dense, so its composition is quite

Composite picture of Jupiter and its four Galilean satellites assembled from Voyager 1 *photographs. Io (upper left) is nearest to Jupiter, then Europa (centre), Ganymede (below) and Callisto (bottom right).*

THE PLANETS

ABOVE *Jupiter's four giant satellites shown at their correct relative sizes. Ganymede (the largest) is bigger than Mercury, Callisto about the same size, while Io (orange) and Europa (white) are both about the size of our Moon.*

RIGHT *Io—the giant pizza—is the most geologically active body known in the Solar System. No impact craters can be seen because Io is continually being resurfaced by sulphurous lava from its many active volcanoes.*

different, probably a rocky core surrounded by a deep ocean of water with a thick crust of rock and ice on the outside. The fact that the craters are so well preserved shows that there are no geological processes (earthquakes, lava-flows, erosion) to destroy them: Callisto is a dead world.

Ganymede is the largest of Jupiter's satellites but only a little bigger than Callisto, so it might have been expected to look much the same; but it does not. Ganymede does have craters, many of them with white haloes which probably consist of water ice that was splashed out by the impacts, but there are not nearly as many as on Callisto. Also, there are many signs of geological activity, such as mountains, ridges, complex grooved structures and light and dark bands. Most exciting of all for the geologists, there is evidence of surface movement similar to continental drift on Earth. No one knows why Ganymede is active and Callisto is not, but it may have something to do with the energy provided by tidal interaction with the other satellites. More obvious indications of tidal interaction are that the orbital period of Ganymede is twice that of Europa and four times that of Io, and that all four Galilean satellites have been *despun* so that they always present the same face to Jupiter, just as our Moon does to Earth.

Europa is yet another surprise. Its size and density suggest a similar structure to the Moon, and so it probably is inside. But its surface is much more like that of an ivory billiard ball. It is very smooth, but criss-crossed with darker lines and with some curious scalloped lighter lines. The surface is thought to be a layer of ice, but the lines remain a mystery.

Io is the innermost Galilean satellite and even before the *Voyagers* it was known to be odd. Astronomers had found it to have red polar caps, to influence radio emission from Jupiter and to have a very thin atmosphere containing, to their surprise, sodium. Thus the *Voyager* pictures were awaited with great interest, but nobody could have anticipated what they showed. Io looks just

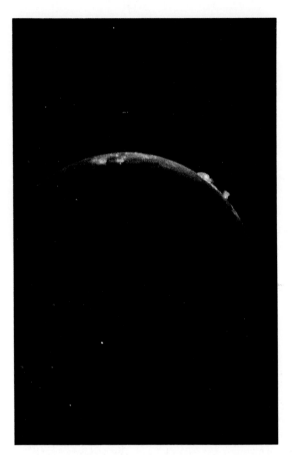

Voyager 2 *monitored Io's volcanic activity continuously for six hours. In this picture, two blue volcanic plumes have been ejected to a height of about 100 kilometres above the surface.*

like a giant pizza. There is no sign of any impact craters, though Io must have been bombarded at least as heavily as the other satellites, so there must be some active geological processes at work to wipe them out. That this is indeed so was demonstrated by pictures showing several volcanoes erupting, and erupting much more violently than anything ever seen on Earth. The clearest evidence for the eruptions was the plumes of ejecta seen against the sky at the edge of Io's disc. The volcanic material (sulphur and sulphur dioxide) is thrown out to heights of over 200 kilometres, and then rains back onto Io, the sulphur dioxide as white crystals and the sulphur giving the brilliant red, orange and yellow colours. Most of the surface is covered in these chemicals, produced from volcanic outbursts or lava-flows. It seems that there is a sea of molten sulphur just below Io's crust which is heated and squeezed outwards by the tidal interactions between Io, Jupiter and the other satellites.

As if sixteen satellites were not enough, Jupiter also has a ring! This was discovered by the *Voyagers* and shows up best when viewed from the shadow of Jupiter looking back nearly in the direction of the Sun. It is a pretty feeble ring, though: nothing like the magnificent ring structure around Saturn.

Jupiter's ring, discovered by Voyager 1 *and photographed here by* Voyager 2. *Such a clear view is possible only from beyond Jupiter, looking back towards the Sun.*

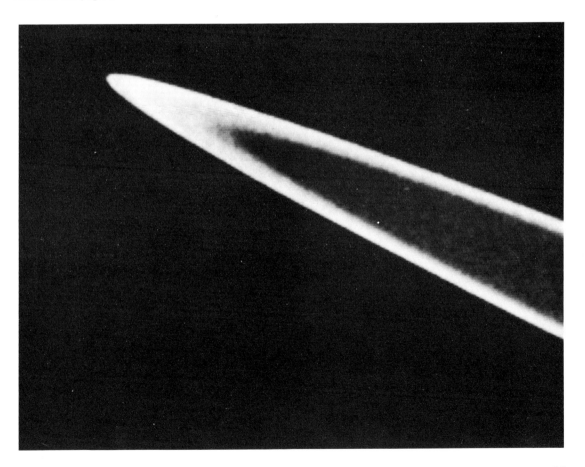

8 · Saturn

Orbital period: 29.46 years; *rotation period*: 10.66 hours; *distance from Sun*: 1507×10^6 km (maximum), 1347×10^6 km (minimum); *diameter*: 120×10^3 km (equatorial), 108×10^3 km (polar); *mass*: 5.69×10^{26} kg; *density*: 0.7×10^3 kg m^{-3}.

Satellites

	Name	Orbital period	Radius of orbit (10^3 km)	Diameter (km)	Mass (10^{18} kg)
XV	Atlas	14.45 hr	138	40×20	
	1980S27†	14.71 hr	139	$140 \times 100 \times 80$	
	1980S26†	15.98 hr	142	$110 \times 90 \times 70$	
XI	Epimetheus	16.66 hr	151	$140 \times 120 \times 100$	
X	Janus	16.69 hr	151	$220 \times 200 \times 160$	
I	Mimas	22.62 hr	186	392	45
II	Enceladus	1.370 d.	238	500	74
III	Tethys	1.888 d.	295	1,060	740
XIII	Telesto	1.888 d.	295	$34 \times 28 \times 26$	
XIV	Calypso	1.888 d.	295	$34 \times 22 \times 22$	
IV	Dione	2.737 d.	377	1,120	1,050
XII	1980S6	2.737 d.	377	$36 \times 32 \times 30$	
V	Rhea	4.518 d.	527	1,530	2,500
VI	Titan	15.95 d.	1,222	5,150	135,000
VII	Hyperion	21.28 d.	1,481	$410 \times 260 \times 220$	17
VIII	Iapetus	79.33 d.	3,561	1,460	1,900
IX	Phoebe	1.51 yr*	12,952	220	0.4

*Retrograde orbit.
†Provisional designation. These satellites have yet to be named and numbered.

The view from Earth

Saturn is the most distant of the planets known to the ancients, orbiting the Sun at nearly twice the distance of Jupiter. It is very easy to see with the naked eye, but you need to know roughly where to look as it is no brighter than some of the brightest stars and does not move noticeably from night to night.

It is only when the planet is viewed through a telescope or binoculars that the truly remarkable

The planet Saturn.

nature of Saturn is revealed. Inevitably, Galileo was the first to try this (in 1610) and he realized at once that he was seeing something peculiar. He wrote to the Grand Duke of Tuscany 'the planet Saturn is not one alone but is composed of three, which almost touch one another ... the middle one is about three times the size of the lateral ones'. Two years later he began to doubt his telescope, if not his reason, because the two smaller companions had disappeared. However, they were visible again a year later. It was not until 1655 that the true nature of the 'companions' was discovered, when the young Dutch scientist, Christian Huygens, examined Saturn through a much better telescope and realized that what he was seeing was a planet with a ring round it. The ring is only slightly inclined to the line of sight from Earth, and the inclination varies according to the relative positions of the Earth and Saturn in their orbits. Galileo had first seen the ring two years before the inclination decreased to zero and his subsequent failure to observe it was because it was then edge-on to the line of sight from Earth. The ring is so thin that it virtually disappears when this happens.

If you have a telescope or binoculars, they are almost certainly of better quality than Galileo's telescope and quite capable of showing you Saturn and its rings, but remember that you have the advantage of knowing what you are looking for whereas it was all completely new to Galileo. Even with the biggest telescope the view is grossly inferior to the familiar spacecraft pictures, but there is absolutely no substitute for seeing the rings for yourself, however imperfectly. This is the most remarkable sight the heavens have to offer.

Saturn and its satellites

Apart from its ring system, Saturn is in almost every respect a slightly inferior version of Jupiter. It has the same composition (mostly hydrogen and helium) and a similar structure, though the metallic hydrogen core is relatively smaller because Saturn is a smaller planet and the pressure required to convert hydrogen to its metallic form is only found deep within it. It has a similar atmosphere with bands of clouds parallel to the equator, but this is less turbulent than Jupiter's and shows only a small brown spot instead of a great red one. Saturn has more satellites than Jupiter—the present count is 17—but only one of these, Titan, is comparable in size with Jupiter's four Galilean satellites. Although Titan is larger than Mercury and is of special interest because it has an atmosphere, it still has to take second place to Jupiter's Ganymede in terms of size and mass. Whereas the outermost four of Jupiter's satellites have retrograde motions, only one of Saturn's satellites, Phoebe, is retrograde.

The satellites we have looked at so far have all been either very large (the Moon, Jupiter's Galilean satellites, Titan) or very small (Phobos, Deimos and the remainder of Jupiter's satellites). But Saturn has six of intermediate size, from 400 to 1500 kilometres in diameter. They are believed to be composed mainly of ice, but with a little rocky material present as well. The outer two are Rhea and Iapetus, both about 1500 kilometres in diameter, with one inside and one outside the orbit of Titan. Both are heavily cratered, but whereas Rhea is uniformly bright, Iapetus has one bright side and one as black as tar. The black side is the one that faces the direction in which Iapetus is travelling, and it is tempting to suppose that it is covered with dirt that has been swept up by the satellite as it moves along its orbit. However, the insides of some of the craters on the trailing side have a similar dark coating, which leads some scientists to believe that it comes from within Iapetus. Careful inspection shows there is also a difference between the leading and trailing sides of Rhea: the trailing side has bright wispy markings superimposed on the cratered surface.

Dione and Tethys are closer to Saturn, both about 1100 kilometres in diameter and heavily cratered. Tethys has the largest crater of the Saturn system—a huge impact crater 400 kilometres across. Dione has the same bright

Our view of Saturn's rings depends on the position of the planet in its orbit. At (1) they appear tilted down, at (3) they appear tilted up and at (2) and (4) they are seen edge-on and virtually disappear.

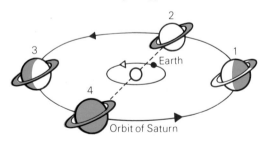

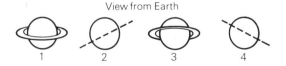

Composite picture of Saturn and its six largest satellites assembled from Voyager *images. Moving clockwise from Dione (foreground), they are Enceladus, Rhea (bottom left), Titan (top left), Mimas and Tethys.*

The brilliant white surface of Enceladus, one of Saturn's icy moons. The absence of craters from large regions indicates that the surface there has been renewed.

wispy markings on its trailing side as Rhea, but otherwise its trailing side is darker than the leading side. The two icy satellites closest to Saturn are Enceladus and Mimas. Enceladus appears to be geologically active since the scars of old craters have been completely wiped out over much of the surface, probably by the eruption of water volcanoes, though no active ones have been seen. The fact that Enceladus's surface is extremely bright and whiter than fresh snow is further evidence that the surface is being renewed. The energy for this geological activity probably comes from tidal interactions with the other satellites, but the details have yet to be worked out. In contrast, Mimas is totally inactive and so densely covered with craters that a new one would have to obliterate an old one. Its most striking feature is the crater Arthur, which is about a third of the satellite's diameter. Had the impact that caused it been only slightly greater, the satellite would have broken up.

The remaining Saturnian satellites are all rather small, but some of them have interesting orbits. The outermost is Phoebe, which goes round Saturn the wrong way. Phoebe may have joined Saturn more recently than the original family of satellites. It is extremely dark, and it may be dust from Phoebe that has blackened the leading side of Iapetus. Two small satellites share the same orbit as Tethys, one being a sixth of an orbit ahead of Tethys and the other the same distance behind. The places they occupy are known as *Lagrangian points* after the eighteenth-century French mathematician who showed that the combined gravitational forces of a planet and a large satellite would keep small satellites firmly in their place at these points. All the larger satellites have Lagrangian points, and another small satellite has been found in one of those associated with Dione.

Closer in to Saturn, only just outside the rings, are two more small satellites that share nearly the same orbit. Both orbit Saturn in about 17 hours, but one gains slightly on the other and catches up every four years. When this happens, just as the satellites are about to collide, the gravitational forces between them cause them to swap orbits so that the leading one gets a push that sends it off on a four-year chase of what has now become the slower one. There are three more small satellites even closer in. These are called 'shepherd satellites' because their gravitational influence helps to keep the material that makes up the rings in tidy orbits, just as a sheepdog keeps a flock together.

Titan is Saturn's one monster satellite, being larger than the Moon and Mercury and only slightly smaller than Ganymede. It is just about the same size and density as Jupiter's Ganymede and Callisto, and probably has much the same structure: a rocky core overlaid by a thick coating of ice and topped by a 100-kilometre layer of rock and ice. Titan is not much to look at, just a featureless dull red ball, but this is because it is unique among the satellites in having a proper atmosphere. All that shows up on photographs is the top of the uniform smog layer that shrouds it. The atmosphere is mainly composed of nitrogen (like our own) and methane, but smaller quantities of a number of compounds have been detected, including those that form the essential building blocks of living matter. It would be quite unjustified to assume that there is, or ever has been, any form of life on Titan, but its atmosphere provides an interesting example of what conditions might have been like on Earth before life appeared. In many ways, Titan is more Earth-like than anywhere else in the Solar System, but it is very much colder with a surface temperature of about $-180°$C.

Saturn's satellite Mimas must have come close to being shattered by the impact that caused its giant crater, Arthur.

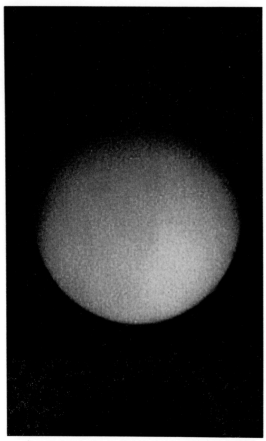

Saturn's Titan is the second largest satellite in the Solar System (after Jupiter's Ganymede) and is larger than the planet Mercury. It is the only satellite to have a substantial atmosphere.

Finally we come to Saturn's rings. Even before close-up pictures were obtained from spacecraft, the rings had been extensively studied by many famous astronomers. As early as 1676 Jean Cassini observed that there are at least two rings, separated by a clear division which is named in his honour. He also suggested that the rings are not solid, but composed of many tiny satellites and this was confirmed by James Clerk-Maxwell who showed mathematically that a solid disc would break up. The Cassini division can be seen with quite a small telescope, but it takes something rather more powerful to show up a faint inner ring that was discovered in 1850 and called the crêpe ring because it appears so flimsy.

One of the questions about Saturn's rings is: how thick are they? The two outer rings look pretty substantial and they cast strong shadows on Saturn but, as Galileo was disconcerted to find, the rings almost disappear when seen edge-on. The rings are incredibly thin relative to their width, just a few kilometres thick. In 1848 Edouard Roche explained why the rings should be there at all. He showed that tidal forces will break a satellite up if it gets too close to a planet. Too close was calculated to be within 2.4 times the radius of Saturn (called the Roche limit), and it is just within this distance that Saturn ceases to have satellites and starts to have rings. The rings are composed of millions of chunks of ice, kept in their place by shepherding satellites and by resonances with the others. The Cassini division is an example of resonance. Any lump of ring material orbiting there would have a period half that of Mimas and a third that of Enceladus, and would receive gravitational pulls from these satellites which would move it to another orbit clear of the division.

It was anticipated that other satellites would produce other resonances which would give some structure to the rings, but no one anticipated the incredible amount of fine structure shown by the *Voyager* photographs. These showed the rings to be as heavily banded as a record disc, with many additional curious features, such as spokes, braids and twisted rings. When Galileo had his first indistinct view of the rings, little did he realize what a truly magnificent structure they really are.

SATURN

This high-resolution picture taken by Voyager 2 *shows the extremely complex structure of Saturn's rings. A number of 'spokes' can be seen as well as innumerable concentric rings.*

9 · Uranus, Neptune and Pluto

All the planets discussed so far are easily seen with the naked eye and have been well known from ancient times. The outer three planets (Uranus, Neptune and Pluto) have been discovered only with the aid of telescopes.

Uranus

Orbital period: 84.01 years; *rotation period*: 17.3 hours; *distance from Sun*: 3007×10^6 km (maximum), 2737×10^6 km (minimum); *diameter*: 51×10^3 km; *mass*: 8.70×10^{25} kg; *density*: 1.23×10^3 kg m^{-3}.

Satellites

	Name	Orbital period	Radius of orbit (10^3 km)	Diameter (km)	Mass (10^{18} kg)
	1986U7	7.9 hr	49	15	
	1986U8	8.9 hr	53	20	
	1986U9	10.4 hr	59	50	
	1986U3	11.1 hr	62	70	
	1986U6	11.4 hr	63	50	
	1986U2	11.8 hr	64	70	
	1986U1	12.3 hr	66	90	
	1986U4	13.4 hr	70	50	
	1986U5	14.9 hr	75	50	
	1985U1	18.2 hr	86	170	
V	Miranda	1.41 d.	129	480	180
I	Ariel	2.52 d.	191	1,170	1,600
II	Umbriel	4.14 d.	266	1,190	1,000
III	Titania	8.71 d.	436	1,590	5,900
IV	Oberon	13.46 d.	583	1,550	6,000

In 1756 William Herschel came to England from Hanover as a boy musician and eventually became organist at the Octagon Chapel in Bath. His interest in astronomy was kindled by Robert Smith's book on optics which he read because the same author had also written a book on

URANUS, NEPTURE AND PLUTO

Sir William Herschel, who discovered the planet Uranus in 1781.

harmonics. As a result of this Herschel started to build reflecting telescopes and went on to become the greatest telescope-maker of his time. He not only built telescopes, but he also used them for systematic surveys of the sky, discovering and cataloguing double stars and galaxies. On 13 March 1781, he came across an object that was clearly not a star or a galaxy and was certainly not one of the known planets. He first thought that he had discovered a comet and he announced this discovery to the Bath Philosophical Society and also communicated it to the Royal Society of London. When the orbit was determined it was found not to be a comet; Herschel had become the first man ever to discover a planet. He named it *Georgium Sidus* in honour of King George III, but the name was later changed to Uranus.

It was subsequently found that earlier astronomers had observed Uranus, but none of them had realized that it was not a star. Herschel became a full-time astronomer and spent the rest of his life observing and making huge telescopes. Among his many achievements was the discovery of two of Saturn's satellites (Mimas and Enceladus) and the accurate determination of Saturn's period of rotation.

The view from Earth

Uranus is the seventh planet out from the Sun. In principle, it can just be seen with the naked eye, but it looks no different from a faint star so it is almost impossible to identify it. Even with a small telescope it is necessary to have a good finder-chart to help distinguish it from the background stars. The British Astronomical Association includes such a chart in their annual handbook. Uranus shows a distinct greenish disc through a large telescope, but there are no noticeable markings so the rotation period had to be estimated by another method. The curious thing about the spin of Uranus is that the planet's axis of rotation is nearly in the plane of the orbit instead of (like all the other planets) almost at

Because Uranus's axis of rotation is nearly in the plane of its orbit, the Sun sometimes shines steadily over the planet's north pole, sometimes over the south pole and at other times rises and sets over the equator.

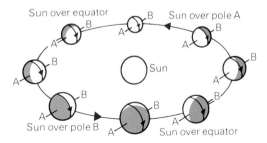

right-angles to it. This means that at one time during its year the Sun appears steadily over the north pole of Uranus. For the next 42 years it then spirals slowly downwards towards the equator and on towards the south pole, when the north pole will spend several years in darkness. The Sun then spirals back towards the north until, 84 years later, it gets back to where it started from. The view from the Earth is not much different to that from the Sun, so we sometimes see only the northern hemisphere of Uranus, sometimes only the southern hemisphere, as at present, and in between get a view from above its equator.

Five satellites of Uranus have been detected from earth-based observations, and they have been given rather pleasing names from Shakespeare and Alexander Pope as a change from classical mythology; starting from the outermost they are Oberon, Titania, Umbriel, Ariel and Miranda. They all orbit above Uranus's equator, so we sometimes see the orbits from above, sometimes from below and sometimes edge-on.

A most remarkable discovery about Uranus was made in 1977 as the result of an attempt to learn something about the planet's atmosphere by studying the fading of a star as Uranus passed in front of it. However, shortly before the star was eclipsed it was found to fade and then brighten up again several times. A matching set of fadings was noted after the eclipse. The conclusion was clear—Uranus has rings! The star's light faded as it passed behind each ring. These rings are not bright and wide like those of Saturn, but narrow dark bands. This was the first indication that planets other than Saturn have got rings, as the ring around Jupiter was not discovered until later. There is now some evidence that Neptune, too, has a ring, so perhaps the Earth should now be thought of as rather peculiar in *not* having one.

Spacecraft exploration

This was about as much as could be said about Uranus until it was visited by *Voyager 2* early in 1986. As we have seen, *Voyager 2* was launched back in 1977, encountered Jupiter in 1979 and Saturn in 1981, then continued to Uranus. Its work is still not finished: it is on course for an encounter with Neptune in 1989—not a bad record for one small spacecraft.

The *Voyager* photographs of Uranus were, frankly, disappointing, showing only the mildest of variations on the cloud-covered surface. Uranus was found to have a magnetic field and, in the absence of surface markings, the rate of rotation of the magnetic field gives the best estimate of the planet's rotation—just over 17 hours. Since it has a magnetic field, Uranus must have a conducting interior, but what is it made of? The present view is that it has a rocky core of 10,000 kilometres diameter inside a mantle of hot, dirty water; it is the pollution in the water that makes it a conductor. The atmosphere of Uranus is composed of hydrogen and helium in

Even at its closest approach, Voyager 2 was unable to resolve any obvious structure on the uniformly cloud-covered surface of Uranus.

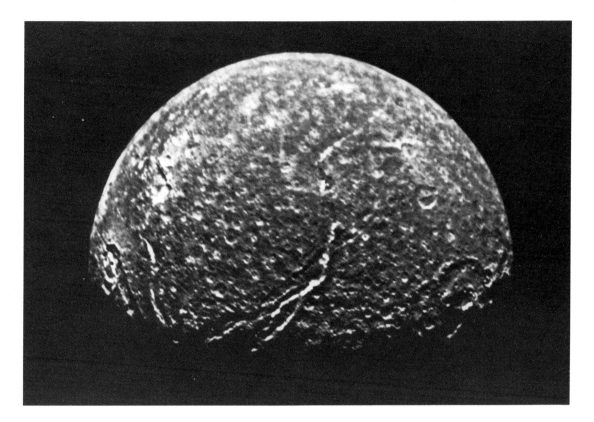

Titania, the largest of Uranus's satellites, is uniformly covered by impact craters. The long rifts may be cracks caused by water inside the satellite freezing and expanding.

about the same proportions as in the atmospheres of Jupiter and Saturn, together with the clouds of methane which give Uranus its characteristic blue-green colour.

If the pictures of Uranus itself were dull, those of the satellites and rings more than compensated. The five satellites known before the *Voyager* encounter are all ice worlds, like the medium-sized satellites of Saturn, with grey surfaces. The surface of the outermost, Oberon, is spotted with craters with radiating spikes of lighter material, probably cleaner ice. The splodges of dark material inside some of the craters are rather more puzzling; could this be a sign of water-volcano activity, with dirty water penetrating up from the interior to fill the craters?

Oberon's other surprise was a 5-kilometre high mountain (massive for such a small satellite), that stood out from the disc against the dark sky background.

The next satellite, Titania, is the largest. This showed many craters, up to 400 kilometres across, and some huge canyons that may have been caused by surface expansion as a result of water in the interior freezing. In contrast, in all ways but one Umbriel appears to be long dead, its crater scars all uniformly covered with a grey

THE PLANETS

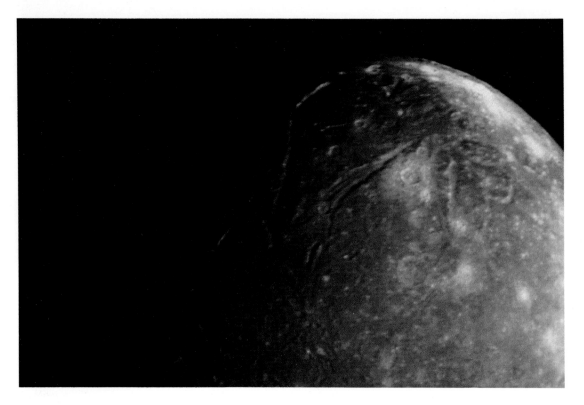

The most obvious features on Ariel are the bright patches associated with craters, but the flat-bottomed valleys are equally interesting. These have very few craters in them, implying that the in-fill is not the original surface.

dust that makes it the darkest of all Uranus's satellites. The one exception is a large, bright, doughnut-shaped ring which has so far defied explanation. Ariel, the next satellite in, is just about the same size and mass as Umbriel but there the similarity ends. Ariel has light and dark areas and a wealth of criss-crossing canyons as well as smooth-bottomed valleys. Miranda is probably the most curious of all the satellites in the Solar System. Its surface is divided up into well-defined regions, each of which has an entirely different terrain from the others. One is a bulls-eye of concentric ridges, one is cratered, another is a darker region with a distinctive light chevron in the middle. In addition there is a long jagged cliff 5 kilometres high and so on. It looks almost as if Miranda has been made up of all the bits left over from the rest of the Solar System. Indeed, it has been seriously suggested that Miranda was broken to pieces by a huge meteorite impact in the distant past and that the

Miranda, a most peculiar satellite which may be a jigsaw of pieces from an earlier satellite that got broken. The apparent elongation of Miranda towards the 5-kilometre high cliff is not real, but results from the changing position of Voyager 2 during the course of the imaging.

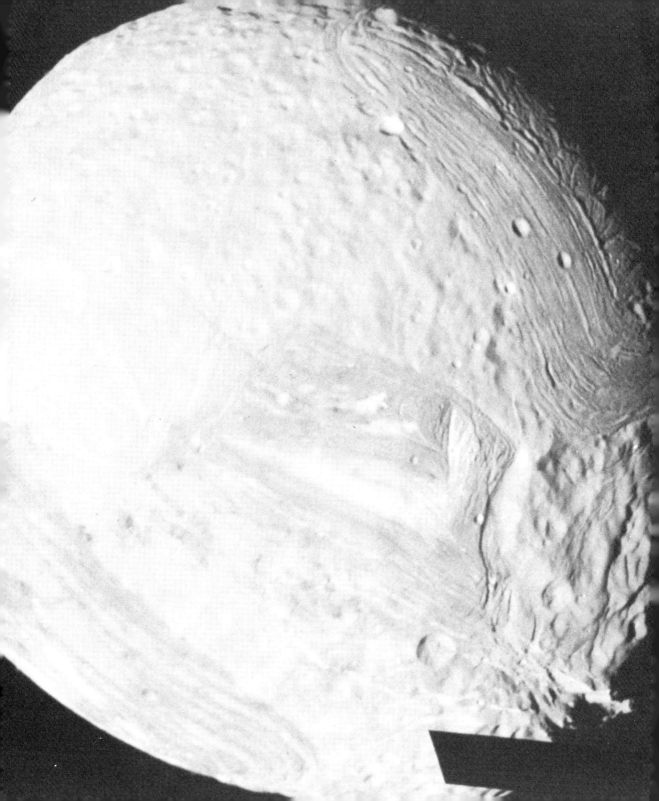

Uranus's rings can best be seen looking back towards the Sun from beyond the planet. The trailed images are background stars.

present patchwork is the result of the separate pieces getting together again.

In addition to taking pictures of these five larger satellites, *Voyager* discovered a further ten small ones, several of which are 'shepherds', keeping the rings in order. As expected, the rings are narrow bands of very dark material and are best seen when looking back towards the Sun. We cannot do this from Earth, since Uranus is further from the Sun than we are, but *Voyager* went beyond Uranus and was able to look back at the rings. They were seen to be much more complicated than the ten-band structure seen from the sunward side, but not in the same class as Saturn's rings. Radio observations showed that the rings are composed of lumps a metre or more across, with hardly any smaller particles, quite unlike the rings of Saturn and Jupiter.

Neptune

Orbital period: 164.8 years; *rotation period*: 15.8 hours; *distance from Sun*: 4540×10^6 km (maximum), 4462×10^6 km (minimum); *diameter*: 51×10^3 km; *mass*: 1.03×10^{26} kg; *density*: 1.66×10^3 kg m^{-3}.

Satellites

Name	Orbital period (days)	Radius of orbit (10^3 km)	Diameter (km)	Mass (10^{18} kg)
I Triton	5.876*	354	3,800	140,000
II Nereid	360.2	5,511	300	20

*Retrograde orbit.

Without wishing to detract from Herschel's achievement in discovering Uranus, it has to be admitted that it was largely a matter of luck, aided by the fact that Uranus is reasonably bright. In contrast, the discovery of Neptune was the result of very careful scientific detective work although there was also an element of luck involved.

Following the discovery of Uranus, careful measurements were made of its changing position in the sky. The planet's orbit was calculated from these observations and, as expected, it was found to be an ellipse round the Sun. However, it was not the perfect ellipse that it would have been if the Sun and Uranus were the only objects in the Solar System: there were small perturbations due to the gravitational attraction of the other planets, in particular Jupiter and Saturn. Knowing the masses and positions of the planets, it is a difficult but not impossible task to calculate how they would modify the orbit of Uranus, and this was undertaken in 1843 by a brilliant young Cambridge mathematician, John Couch Adams. After much painstaking work, he was able to account for most of the perturbations, but not all of them.

This was when Adams's true genius came into play, first in inferring that this must be due to the presence of some as yet unknown planet, and then in undertaking the Herculean task of deducing the mass and orbit of the planet from the tiny differences between the theoretical and observed orbit of Uranus. When he eventually achieved this, and calculated precisely where in the sky the new planet should be, the obvious next stage was to look for it. Since he was not an observer, Adams sent his predictions to the Astronomer Royal, George Airy, at Greenwich. Airy dismissed Adams's work without bothering to point a telescope to the position indicated because he believed the perturbations to be due to another cause. He later forwarded the information to the Cambridge astronomers, who started a leisurely but fruitless search.

Meanwhile another gifted mathematician, Urbain Le Verrier, had been addressing the same problem in France, quite independently of Adams. By June 1846 he, too, had calculated a position for the supposed new planet remarkably similar to that obtained in 1845 by Adams, though neither of them was aware of the other's work. Le Verrier sent his predicted position to the Berlin Observatory, where Johann Galle saw and identified the planet the same night he received the information, because it did not appear on a new and accurate star map that he had recently obtained. The discovery was confirmed the following night (24 September 1846), when the new object was found to have moved slightly relative to the stars.

So Neptune had been discovered, but by whom? Galle had been the first to see and recognize it (it later transpired that it had been seen but not recognized during the Cambridge search), but he could not have done so without Le Verrier's predictions. Should the credit go to the man who predicted it, in which case Adams would have the prior claim, even though his predictions were not used? Posterity has awarded equal credit to Adams and Le Verrier, with a 'highly commended' to Galle.

Neptune has not yet had the privilege of a visit by a spacecraft, but *Voyager 2* is on its way to an encounter in 1989. Meanwhile, nearly all our knowledge of Neptune is based on observations made from Earth. It is too faint to be seen with the

Neptune and its larger satellite, Triton, photographed by Gerard Kuiper with the 82-inch telescope at the McDonald Observatory in Texas.

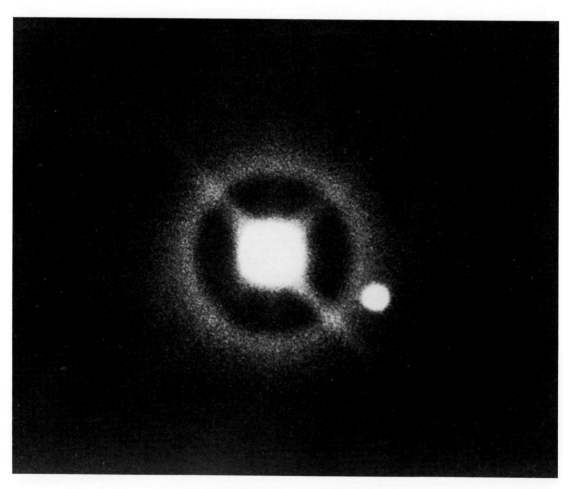

naked eye, but is not too difficult to find with a small telescope or binoculars, though you will need an accurate chart of its position amongst the nearby stars to find it, as it looks just like a star. It can be seen as a disc only through a large telescope on a good night, because its diameter in the sky is no greater than that of a 1p coin seen from a distance of two kilometres. It has a slightly bluish appearance, with no visible surface markings, probably because it is cloud covered.

From its size and appearance, Neptune would appear to be almost an exact copy of Uranus, but after the surprises sprung on us from close-up examination of the other planets and satellites, the only safe policy is to wait and see. Two Neptunian satellites have been discovered. Triton was found by William Lassell within a month of Neptune's discovery, and Nereid by Gerard Kuiper in 1949. Triton is a giant satellite, possibly as large as Titan or Ganymede, though present measurements are somewhat uncertain. It completes a circular but retrograde orbit in under six days, and calculations suggest that it is getting closer to Neptune so that its ultimate fate may be to break up and form a ring. Nereid is a more modest satellite in a safe orbit, but the orbit is very elongated and is a long way from the plane of Neptune's equator. Recent observations of the fading of stars as Neptune passes in front of them suggest that the planet has a ring around it, though this may be a partial ring rather than a complete one.

Pluto

Orbital period: 248.5 years; *rotation period*: 6.4 days; *distance from Sun*: 7388×10^6 km (maximum), 4442×10^6 km (minimum); *diameter*: 2200 km; *mass*: 1.3×10^{22} kg; *density*: 2.1×10^3 kg m^{-3}.

Satellite: Charon; *orbital period*: 6.39 days; *radius of orbit*: 19×10^3 km; *diameter*: 1160 km; *mass*: 1×10^{21} kg?

The most remarkable thing about Pluto is that it was discovered at all. The procedure was essentially the same as that which led to the discovery of Neptune, though much more systematic.

The orbit of Neptune, like that of Uranus, was found to depart slightly from what could be accounted for by the known members of the Solar System. Following the example of Adams and Le Verrier, Percival Lowell and William Pickering (two American astronomers) assumed that the departures were due to the influence of some yet-more-remote unknown planet, and attempted to calculate where it would be in the sky. The predicted position was less precise than that for Neptune, and the new planet was expected to be much fainter, so it was necessary to conduct a very careful systematic search. A telescope was installed at the Lowell Observatory in Arizona to carry out this search. The plan was to take a set of photographs of the appropriate part of the sky, then take a repeat set sometime later, and compare the two sets to see if any of the images had moved. The search took years, but in 1931 Clyde Tombaugh found what he was hunting for within a few degrees of the predicted position.

THE PLANETS

LEFT *Clyde Tombaugh, who discovered the planet Pluto in 1931, photographed by Patrick Moore in 1980.*

The new planet was named Pluto.

But is Pluto the planet that was predicted by Lowell and Pickering? To produce the observed perturbations to Neptune's orbit, it would need to be about the same mass as the Earth, whereas all the measurements of Pluto's diameter made at that time indicated that it was less than half that of the Earth. To pack the same mass into such a small planet would require it to be made of some material ten times as dense as the Earth, which seems unlikely. It now seems that Pluto is smaller than our Moon and about one-sixth of its mass— much too small to be the object predicted. Tombaugh's discovery was made after a painstaking search and can hardly be dismissed as sheer chance, but it does appear fortuitous that the planet he discovered was near to the predicted position. That raises the interesting possibility that there may be yet another planet (or planets) beyond Neptune.

Despite its small size, Pluto does have a satellite, and quite a respectable one. Charon has half the diameter of Pluto and about a tenth of its mass, though this value is uncertain. It rotates around Pluto in the same time that Pluto rotates, so Charon and Pluto always face one another like a pair of Viennese waltzers. Pluto is not always the outermost planet of the Solar System. This is because its orbit is highly eccentric and part of it is closer to the Sun than Neptune's. Pluto is in that part of its orbit now and will remain there until 1999. An ingenious but highly speculative explanation of the curious orbits of Pluto and Neptune's satellites suggests that Pluto was a satellite of Neptune until a close encounter with Triton threw it clear and at the same time reversed the direction of Triton's orbit.

RIGHT *Pluto looks just like a rather faint star. It is its day-to-day movement relative to the background stars that shows it to be a planet.*

URANUS, NEPTURE AND PLUTO

10 · Interplanetary Debris

As well as the planets, there are many other small bodies orbiting the Sun. These include asteroids, comets, meteors, and meteorites.

The asteroids

A numerical device known as Bode's law gives the approximate distances of the planets from the Sun. It was devised by Johann Titius and publicized by Johann Bode (both German) in 1772. The procedure is as follows: add 4 to each member of the series 0, 3, 6, 12, 24 ... and divide by 10. This gives:

| 0.4 | 0.7 | 1.0 | 1.6 | 2.8 | 5.2 | 10.0 | 19.6 |

compared with the real values (in a.u.)

| 0.4 | 0.7 | 1.0 | 1.5 | — | 5.2 | 9.5 | 19.2 |

for the planets

| Mercury | Venus | Earth | Mars | ? | Jupiter | Saturn | Uranus |

The initial series is a bit of a fudge because it is obtained by doubling the previous number and should really start with 1.5, not zero. Nevertheless, it gives a surprisingly good fit not only to the six planets known in 1772 but also to Uranus, which was not discovered until 1781 (it breaks down badly for Neptune and Pluto). The odd thing is the gap between Mars and Jupiter where Bode's law suggests there should be another planet. Although it was merely 'playing with numbers' and had no scientific basis, Bode's law was so good for the known planets that several astronomers thought it worthwhile to look for a planet at 2.8 a.u.

Nothing was found until 1801 when the Italian astronomer Giuseppe Piazzi discovered Ceres at 2.76 a.u. from the Sun. But Ceres, with a diameter of only about 1000 kilometres could hardly be called a planet. It proved to be the first and largest of many such objects to be discovered in the gap between Mars and Jupiter. They are known as *asteroids*; most of them are less than 100 kilometres across and there are tens of thousands of them. Most of the asteroids have fairly circular orbits with radii within 0.5 a.u. of the Bode's law prediction. They might be thought of as another example of a ring system, though a very sparse one and certainly not the result of tidal forces within the Roche limit. Just as Mimas and Enceladus are responsible, through resonance, for the Cassini division in Saturn's rings, so Jupiter causes gaps in the asteroid belt—no asteroids have been found with orbital periods that are $\frac{1}{3}$, $\frac{1}{2}$, $\frac{2}{5}$, or $\frac{3}{7}$ that of Jupiter. Jupiter has also collected some of the asteroids into its Lagrangian points (see p. 71), where they share the same orbit as Jupiter, but lead or lag the planet by one-sixth of an orbit. These are called the *Trojan asteroids*.

Some of the asteroids have more elliptical orbits, probably as a result of close encounters or collisions with other asteroids in the past. These take them outside the orbit of Jupiter or inside the orbit of Mars and a few of them even get closer to the Sun than the Earth. The first asteroid found to have such a highly elliptical orbit was called Apollo, and the ones discovered subsequently are known as Apollo asteroids, though they also have names of their own such as Eros and Icarus.

As we have seen (Chapter 1), Eros proved particularly useful in the measurement of the scale of the Solar System because it came closer to the Earth than any other planetary body—near enough for an accurate measurement of its distance.

A peculiar asteroid-like object with a highly elliptical orbit was discovered by the American astronomer Charles Kowal in 1977 while searching for Trojan asteroids. Known as Chiron, in most respects it looks like just another asteroid, but its orbit takes it from just inside the orbit of Saturn nearly as far out as Uranus, and at no time does it go anywhere near the asteroid belt. Could it be an asteroid that has been deflected from its path by a close planetary encounter, or is it a new type of object?

One of the rewards of asteroid hunting is that the discoverer can recommend a name to the International Astronomical Union, the only body that has the authority to name astronomical objects. (Do not be taken in by the advertisements that offer to put your name to a star—for a price.) One, recently discovered by Dr E. L. Bowell, has been named 'Greenwich'.

Meteorites

The chances of the Earth colliding with an asteroid are very small indeed, but collisions with smaller interplanetary objects are common. The familiar 'shooting stars', streaks of light that flash across the night sky in a fraction of a second, are the result of tiny grains of interplanetary dust being burned up as they enter the atmosphere. Slightly larger lumps, about the size of a pea, can give a more lasting account of themselves, producing a stronger streak that may remain glowing for a second or two before it disappears. Others may produce a fireball—a streak of light ending in a flash as the missile vaporizes and explodes.

Our atmosphere is a good barrier against this heavenly bombardment and most of the material coming into it is completely burned out well above ground level. Occasionally, though, we run into a lump that is large enough to survive its journey through the atmosphere and crash to earth. Such lumps are called *meteorites*. This sounds as though they should be a small meteor, but in fact meteors are shooting stars and much less substantial. Another confusion is that meteorology is the study of the weather and has nothing to do with meteors. So what do you call a person who studies meteors?

Many meteorites have been found, some because they were seen blazing through the sky to the point of impact, but most because they looked odd in their surroundings: a lump of iron in a field, or a rock in a desert, for example. They are not much to look at, but are of particular interest because, until the lunar samples were collected, they were the only specimens of extra-terrestrial material available for study. Meteorites come in many sizes, ranging from pebbles to huge boulders, though fortunately for us there are very few of the larger variety.

What little evidence there is about the origin of meteorites suggests that they may have come from the asteroid belt and are fragments of what was once a much larger body, maybe 500 to 1000 kilometres across. Perhaps the Earth does get hit by asteroids after all, though fortunately it has not yet collided with a really large one.

There have been some sizeable impacts in the past. The most recent is the one which caused the giant Meteor Crater in Arizona. There is also evidence of even larger craters which have been almost weathered away.

Comets

With all the publicity that surrounded the recent reappearance of Halley's Comet and its encounter with the *Giotto* spacecraft, few people can be unfamiliar with what a comet looks like and

THE PLANETS

Meteor Crater, Arizona, a kilometre across and nearly 200 metres deep.

what it is made of. Comets can be most spectacular objects, visible even in daylight and with magnificent tails sweeping half-way across the sky, though this time round Halley's Comet was, for those of us living in England, at best a faint blur in a telescope.

Several comets are discovered each year, but very few of them are bright enough to be seen with the naked eye. Nevertheless, most people will get a chance to see at least one really impressive one sometime during their life. Comets are quite different from asteroids in structure. They are 'dirty snowballs', just a few kilometres across, and can be seen only when they come close enough to the Sun for some of their gases to evaporate and form a reflecting halo round the snowball nucleus. Enthusiastic comet seekers around the world sweep the sky with binoculars each night searching for a tiny, fuzzy blob that does not appear on their star-charts. If they find one, they immediately report it to the International Astronomical Union and when three separate observations have been made an orbit can be calculated. A comet is usually given the names of the first observers to find it.

Most comet orbits are *parabolae*, geometrical shapes that start and finish at an infinite distance so the comets pass close to the Sun only once and then disappear for ever. A few cometary orbits, however, are ellipses, bringing the comet back to the Sun at regular intervals, unless it breaks up or gets perturbed into another orbit by the gravitational attraction of another body, for example, Jupiter. About a hundred such periodic comets have been discovered, by far the most famous of which is Halley. Their orbits are very elongated, with one end near the Sun and the other far out. Like any other orbiting body they have to obey Kepler's laws (Chapter 1), so the Sun is at a focus of the ellipse and they move most rapidly when they are close to it. As an example, Halley's Comet takes seventy-six years to complete an orbit, but is inside the orbit of Saturn for only six of these years.

As a comet approaches the Sun, more and more of the gas in the nucleus evaporates out to form a large but tenuous halo which completely obscures the view of the nucleus itself. The solar wind blows some of the gas away from the comet to form a tail millions of kilometres long, but so

John Emslie's comet and meteor shower drawings are rather fanciful, but the diagram at the centre, illustrating how the comet's tail always points away from the Sun, is sound.

Nucleus of Halley's Comet photographed by the Giotto spacecraft in March 1986.

insubstantial that it is almost a vacuum. The tail always points away from the Sun regardless of which way the comet is travelling, and is visible for only a few weeks near the time of closest approach to the Sun. As the comet moves away again, most of the evaporated material is lost for ever, dispersed in interplanetary space. So each time a periodic comet passes the Sun it loses a bit of its mass, and even the largest cannot be expected to last for more than a few thousand orbits. In terms of the age of the Solar System this is a very short time indeed, so there must be a source of new comets, otherwise they would all have disappeared long ago.

Comets are thought to come from a region of millions of comets in approximately circular orbits somewhere beyond Pluto. This is called the 'Oort Cloud' after the Dutch astronomer who proposed it. Although it cannot be seen, it fits the known facts about comets so well that hardly anyone doubts its existence. The comets we see are those that have been pushed out of the Oort Cloud through close encounters with their fellows, just as the Apollo asteroids were pushed out of the asteroid belt. Since a comet starts out from the Oort Cloud, at some time in its orbit it will return there, and may well suffer a further perturbation. For this reason, the return of a periodic comet can never be certain until it has actually been seen on its way back.

It is not impossible that a comet will one day collide with the Earth; indeed this may already have happened. In 1908 a violent fireball was seen in Tunguska, a remote region in Russia. When the investigators arrived, they found trees thrown to the ground for tens of kilometres around, all pointing away from the centre. Despite intensive searches, no trace of any meteorite material has been found and it is now

Comet Bennett photographed from the Alps in 1970.

thought that the explosion was caused by a collision with a comet. Next time you get a snowball in your face, be grateful it was not the Tunguska one!

A less violent form of encounter occurs when the Earth passes through the orbit of a comet. Comets are messy things and leave bits behind them in their orbits, which show up on Earth as meteor showers. Sometimes hundreds of meteors can be seen in a single night. If their tracks are plotted on a star-chart, they will all be seen to radiate from the same point in the sky. This is because they are all moving in the same direction, and appear to spread out in the same way that beams of sunlight do when seen shining through broken cloud.

The *British Astronomical Association Handbook* gives observing details for meteor showers. Good examples occur near 4 January (the Quadrantids), 6 May (η Aquarids), 13 August (Perseids), 21 October (Orionids) and 14 December (Geminids).

Woodcut of the comet of 1577 by Jiri Daschitzsky. The artist can be seen in the foreground making a sketch, but the finished product still seems to owe a lot to imagination.

11 · Are We Alone?

If the origin of the Solar System is speculative, the origin of life is even more so, but there are some interesting pointers. Firstly, all forms of life (whether humans, fish, plants, trees or viruses) are made up from combinations of the same basic organic molecules—amino acids and nucleotides. Traces of these have been detected in the atmosphere of Titan and also in meteorite material. They can also be manufactured from simple chemicals in the laboratory by making electric sparks in a chamber containing water, methane, ammonia and hydrogen; conditions not unlike those in the atmosphere of the primitive Earth. It is also interesting to look at the most rudimentary form of life, the virus. This can be frozen and crystallized like many other chemicals, but under the right conditions it can reproduce itself. Most important of all, in common with all life forms, it contains DNA (deoxyribonucleic acid). This is a complicated molecule consisting of a string of nucleotides which forms a code that determines whether the cells that reproduce from it will be ants or elephants, viruses or humans. We begin to see the outlines of a sequence that might lead from raw chemicals to life, but there are many gaps and a great deal of conjecture.

Two important questions remain: are there any other planetary systems in the Universe, and is there life elsewhere in the Universe? The answer to the first is almost certainly 'yes'. Although we have yet to observe any planets outside the Solar System, there is indirect evidence of their existence. The strongest evidence comes from data obtained by the infra-red astronomical satellite, IRAS. This detected infra-red radiation, of the type that would be expected from planets, coming from around some nearby stars. Although planets are too cool to give out visible light, they are warm enough to emit infra-red radiation. Also, the minute oscillations of some stars suggest that they have small invisible companions in orbit around them; Barnard's Star, one of our closest neighbours at a distance of 380,000 a.u., is an example.

The question of the existence of extra-terrestrial life has given rise to much discussion, most of which has more to do with emotion than science. One approach is to start with the number of galaxies in the Universe, multiply this by the number of stars in each galaxy, then by the fraction of those expected to have planets, then by the proportion of planets thought to be capable of supporting life, and so on until you end up with a figure representing the number of intelligent civilizations now in existence. This sounds convincing, but one of the factors involved is the proportion of suitable planets that actually evolve life. Since our total knowledge of this is that there is life on Earth (the life-seeking experiments on Mars were inconclusive), the only thing we can say about this factor is that it is not zero. The only honest answer we can give to the question is 'we do not know'.

Index

Reference to illustrations appear in italics.

Adams, John Couch 81–2
Adrastea (satellite of Jupiter) 56
Airy, George 81
albedo (proportion of light reflected) 21, 32, 38
Alexandria 42
Amalthea (satellite of Jupiter) 56
amino acids 93
ammonia 58, 93
Ananke (satellite of Jupiter) 56
Apollo 86
Apollo asteroids 86–7
argon 45
Ariel (satellite of Uranus) 74, 78
Aristarchus 15
Arthur (crater on Mimas) 71, *71*
Ascraeus Mons 55
Aswan 42
asteroids 58, 86–7
astronomical unit 15
 measurement of 38
Atlas (satellite of Saturn) 66
atmosphere 8
 of Earth 45–7
 of Jupiter 58
 of Mars 54, 55
 of Mercury 58
 of Saturn 68
 of Titan 71
 of Uranus 76, 77
 of Venus 39, 40

Barnard's Star 93
Bath Philosophical Society 75
Big Bang 7
Bode, Johann 86
Bode's law 86
Botticelli 39
Bowell, Dr E. L. 87
Brahé, Tycho 13
British Astronomical Association 75, 92

calendar 21
Callisto (satellite of Jupiter) 56, 57, 60, 62, 64, 71
Caloris basin 33–4, *35*
Calypso (satellite of Saturn) 66
carbon dioxide 53, 55
 greenhouse effect 39, 41, 45, 49
Carme (satellite of Jupiter) 56
Cassini, Jean 42
Ceres 86
Charon (satellite of Pluto) 83, 84
Chiron 87
chlorine 39
Clerk-Maxwell, James 72
color, visibility of 21, 48
Comet Bennett *91*
comets 75, 87–92
 comet tail 90
conjunction (inferior and superior) 36–7, 48, *49*
Cook, Captain James 38
Copernican theory 10, *11*
Copernicus, Nicholas (Copernican theory) 10, *12*
craters
 on Ariel 78
 on Callisto 60
 on Deimos 55
 on Earth 87
 on Enceladus 70
 on Iapetus 68
 on Mars 53
 on Mercury 32, *34*
 on Mimas 70
 on Miranda 78, *79*
 on the Moon 21–8
 on Oberon 77
 on Rhea 68
 on Tethys 68
 on Venus 41
crêpe ring 72
cyclone 58

Daschitzsky, Jiri 92
days of the week 8

deferent 10, *11*
Deimos (satellite of Mars) 48, 51–3, 55, 68
density 7, 8
 of Callisto 60
 of Deimos 48
 of Earth 17, 42, 43
 of Europa 64
 of Jupiter 17, 56, 57
 of Mars 17, 48, 53
 of Mercury 17, 30, 32
 of the Moon 42
 of Neptune 17, 81
 of Phobos 48
 of Pluto 17, 83, 84
 of Saturn 17, 66
 of Uranus 17, 74
 of Venus 17, 36, 38
deoxyribonucleic acid 93
Dione (satellite of Saturn) 66, 68, 69, 71

Earth 15, 17, 19, 27, 38, 42–7
 atmosphere 45
 crust 43
 mantle 43
 oceans 45
 plate tectonics 43, 44
eclipses 29, 52
Einstein, Albert 15
Elara (satellite of Jupiter) 56
ellipse 13
epicycles 10, *11*, 13
elongation 30, 36, *37*
Enceladus (satellite of Saturn) 66, 69, 70, 72
Epimetheus (satellite of Saturn) 66
Eratosthenes 42
Eros 32, 87
escape velocity 32
eta Aquarids 92
Etna 66
Europa (satellite of Jupiter) 56, 57, 62, 64
evening star 36

94

INDEX

fireballs 87
fortune telling 9

Galle, Johann 81–2
Galileo 10, *12*, 38, 57, 67
Galilean satellites (Io, Europa,
  Ganymede, Callisto) 57, 59, 61,
  62, 68
Ganymede (satellite of Jupiter) 56, 57,
  62, 64, 71
Geminids 92
geological cycle 45
Georgium Sidus 74
Giotto spacecraft 87
Greenwich 2, 21, 27, 42, 50, 87

Hall, Asaph 52
Halley, Edmond 15, 38
Halley's Comet 40, 87, 88, 89, *90*
helium 57, 59, 60, 68, 76
Herschel, Sir William 74–5
Himalia (satellite of Jupiter) 56
Horrocks, Jeremiah 38
Huygens, Christian 67
hydrogen 7, 8, 19, 57, 68, 76, 93
Hyperion (satellite of Saturn) 66

Iapetus (satellite of Saturn) 66, 68
ice 53, 64, 68, 71, 72, 77
ice-caps 49
icy 70
Io (satellite of Jupiter) 56, 57, 59, 60,
  62, *63*, 64–5
ionosphere 47
infra-red 45, 93
International Astronomical Union 34,
  87, 88
IRAS 93
iron 43
iron oxide 55

Janus (satellite of Saturn) 66
Jupiter 8, 9, 15, 17, 19, 56–65, 68, 86
Jupiter's moons 12

Kepler, Johann 9, 13
Kepler's laws 13, 89
Kowal, Charles 87
Kuiper, Gerard 83

Lagrangian points 71, 86
Lassell, William 83
lava flows 27, 44, 65
Leda (satellite of Jupiter) 56

Le Verrier, Urbain 81–2
life 38, 47, 51, 53, 54, 71, 93
Lowell, Percival 51, 83
Lysithea (satellite of Jupiter) 56

magnetic field 47, 60, 76
magnetosphere 47
maria 21, 33
Mariner spacecraft 32–4, 39, 40, 53,
  55
Mars 9, 15, 17, 48–55
  canali 51
  core 53
  ice-caps 49
  life 53–4
  water 53
Maxwell (feature on Venus) 41, 72
Mercury 9, 10, 15, 17, 19, 30–5, 38,
  57
metallic hydrogen 57, 68
Meteor Crater, Arizona 87, *88*
meteors 87, 92
meteor showers 89, 92
meteorites 27, 87, 93
methane 71, 77, 93
Metis (satellite of Jupiter) 56
metre 43
Mimas (satellite of Saturn) 66, 69, 70,
  71, 72
Miranda (satellite of Uranus) 74, 78,
  79
Moon 8, 9, 10, 15, 21–9, 32, 33
  appearance *22*–6
  history 27, 28
  lunation 21
  man in the 23
  maria 21
  moon rock 29
  terminator 21
morning star 36

Neptune 15, 16, 17, 59, 74, 81–3
Nereid (satellite of Neptune) 81, 83
Newton, Isaac 14, 15
  law of gravity 14, 15, 32, 43
nitrogen 45, 71
nucleotides 93

Oberon (satellite of Uranus) 74, 77
ocean 39, 41, 45, 64
Olympus Mons 53, 54, 55
Oort Cloud 90
opposition 48, *49*
oxygen 45

Pasiphae (satellite of Jupiter) 56
Perseids 92
phases
  of Mercury 30
  of Moon 21
  of Venus 12, 36–8
Phobos (satellite of Mars) 48, 51–3,
  55, 68
Phoebe (satellite of Saturn) 66, 71
Piazzi, Giuseppe 86
Pickering, William 83
Pioneer spacecraft 58, 59
Pioneer Venus 40
planet, meaning of 7
Pluto 15, 16, 17, 19, 74, 83–5
Pole Star 42
Pope, Alexander 76
Ptolemaic theory 10, *11*, 38
Ptolemy 10

Quadrantids 92
quadrature 48, *49*

radar 15, 40
radio-waves 31, 34, 60
red spot (on Jupiter) 58, 59
resonance 72
retrograde motion 10, 48, *49*
retrograde orbit 56, 57, 66, 71, 81
Rhea (satellite of Saturn) 66, 68, 69
rings
  of Jupiter 65
  of Neptune 76, 83
  of Saturn 67–8, 72–3
  of Uranus 76, 80
Roche limit 72
rotation
  of Jupiter 58
  of Mars 51
  of Mercury 31, 32
  of Solar System 8
  of Sun 8
  of Uranus 75, 76
Royal Society of London 75

Saturn 8, 9, 15, 16, 17, 66–73
scarps 33
Schiaparelli, Giovanni 31, 51
seasons 42
Shakespeare, William 34, 76
shepherd satellites 71, 72, 80
shooting stars 87
sidereal period 30
Sinope (satellite of Jupiter) 56

95

Sirius 30
Smith, Robert 74
Solar System 7, 58
spacecraft exploration
   Jupiter 58–65
   Mars 53–5
   Mercury 32–4
   Moon 28
   Saturn 68–73
   Uranus 76–80
   Venus 39–40
stades 43
sulphur 65
sulphur dioxide 65
sulphuric acid 39
Sun 7, 8, 9, 10, 15, 18–21
   formation of 8
   prominences 20
   sunspots 19
   temperature 15
Syene 42
synodic period 30
syzygy 48

Telesto (satellite of Saturn) 66
temperature 8
   of Mars 54
   of Mercury 31
   of Sun 15
   of Titan 71
   of Venus 39
Tethys (satellite of Saturn) 66, 68, 69
Thebe (satellite of Jupiter) 56
thermonuclear reaction 8
Titan (satellite of Saturn) 66, 68, 69, 71, *72*, 93
Titania (satellite of Uranus) 74, 77
Titius, Johann 86
Tombaugh, Clyde 83, *84*
transit
   of Mercury 30
   of Venus 38
Triton (satellite of Neptune) 81, *82*, 83, 84
Trojan satellites 86
Tunguska 90
Tuscany, Grand Duke of 67

Umbriel (satellite of Uranus) 74
ultra-violet 40, 45
Universe, origin of 7
Uranus 15, 16, 17, 59, 74–80
   ring 76
   satellites 74, 76

Vega spacecraft 39, 40
velocity of light 57
Venera spacecraft 39, 40
Venus 9, 10, 12, 15, 17, 32, 36–41
   brightness 37, 38
   phases 12, *12*, *36*, 37–8
Viking spacecraft 53–5
virus 93
volcanoes 44, 53, 64
Voyager spacecraft 58, 59, 60–5, 76–80, 82

water 45, 49, 53, 58, 76
water volcanoes 70, 77
wind terrain 34